Health & Safety at Work

AN ESSENTIAL GUIDE FOR MANAGERS

Ninth edition

Jeremy Stranks

The masculine pronoun has been used throughout this book. This stems from a desire to avoid ugly and cumbersome language, and no discrimination, prejudice or bias is intended.

Originally published as *The Manager's Guide to Health & Safety at Work* in 1990
Second edition 1992
Third edition 1994
Fourth edition 1995
Fifth edition 1997
Sixth edition 2001
Seventh edition 2003
Eighth edition 2006
Revised eighth edition published as *Health and Safety at Work* in 2008
Ninth edition 2010

Kogan Page Limited
120 Pentonville Road
London N1 9JN
United Kingdom
www.koganpage.com

British Library Cataloguing in Publication Data

A CIP record for this book is available from the British Library.

ISBN 978 0 7494 6119 5
E-ISBN 978 0 7494 6120 1

Typeset by Saxon Graphics Ltd, Derby
Printed and bound in India by Replika Press Pvt Ltd

Contents

PART 4: Safety Technology

List of Figures and Tables

FIGURES

TABLES

Preface to First Edition

Health and safety at work is, for many managers, a difficult subject. Apart from being steeped in the law, which can be difficult to interpret, it requires a broad knowledge of many disciplines, such as psychology, engineering, chemistry, ergonomics and medicine, each of which is a subject of study in its own right.

Individual attitudes to health and safety, and indeed the corporate attitudes of organisations, may vary substantially. Is health and safety just a question of complying with the law? Should an organisation that likes to think of itself as caring, promote health and safety as part of that caring philosophy? Or should health and safety be seen as an important feature of the business operation aimed at reducing the losses associated with accidents, ill health, etc?

Increasingly, as we move towards a quality-orientated approach to our business activities, it will be seen that health and safety is an integral feature of such an approach. Everyone has a role to play in the management of quality. Accidents, ill health, sickness absence and damage-producing incidents feature in the 'price of non-conformance' – an important measure in quality management.

How much did accidents and sickness absence cost your organisation last year? What is the cost of your current employer's and occupier's liability insurance? Have you recently been prosecuted and fined for breaches of health and safety legislation? There is no doubt that accidents and sickness represent substantial losses to any company. A meagre 10 per cent reduction in these costs can be significant. A 20 per cent reduction would be marvellous.

There is a need, therefore, for managers to be more knowledgeable about the subject of health and safety from a legal, scientific and technical viewpoint. This book has been written with this objective in mind. I hope it will help managers to understand the subject better and make a contribution to improved levels of health and safety performance.

Jeremy Stranks
1989

Preface to Ninth Edition

In the short time since the revised Eighth Edition was published, two significant statutes have come into operation, namely the Corporate Manslaughter and Corporate Homicide Act, and the Health and Safety (Offences) Act.

The corporate manslaughter legislation has particular significance for those who are 'the controlling mind' (*mens rea*) of an organisation. Senior people, such as chief executives, managing directors, senior partners and 'similar officers of the body corporate' need to be fully aware of their duties under health and safety legislation. They need to ensure that, first, measures for installing and maintaining health and safety procedures are adequately resourced and managed and that second, those providing health and safety-related advice to the organisation are truly competent to do so bearing in mind the risks to which employees and others are exposed while at work. In many cases, senior people will need training to enable them to assess the health and safety performance of the organisation and to perform their duties under the Health and Safety at Work Act and Regulations.

Increasingly, people are beginning to realise that accidents are not solely caused through breaches of the law, however. Accidents are about people, the jobs they do and organisational factors affecting their work. This is why behavioural safety, as a relatively new concept, is gaining considerable significance in the development of strategies for accident prevention. In view of the importance of this subject, the part on behavioural safety has been extended in this edition.

Fundamentally, organisations have got to take health and safety on board and treat it as a significant feature of their activities. The Institute of Directors (IOD) and the Health and Safety Executive (HSE) have recently set out an agenda for the effective leadership of health and safety which senior management should put into practice. Failure to do so could be expensive!

Jeremy Stranks
2010

ACKNOWLEDGEMENT

The author wishes to place on record his thanks to the British Standards Institution.

Extracts from British Standards are reproduced by permission of the British Standards Institution. Complete copies of the documents can be obtained from the British Standards Institution, 389 Chiswick High Road, London W4 4AL.

Part 1

Health and Safety Management and Administration

1

Principal Legal Requirements

The law on health and safety at work has, like much protective legislation, developed in a fragmented way over the last two centuries. In many cases, its growth has been brought about as a result of public outcry at, for instance, the appalling conditions under which children worked in the Lancashire textile mills following the Industrial Revolution. Indeed, the first statute, the Health and Morals of Apprentices Act 1802, was passed to combat this state of affairs in the textile industry. This Act limited the working hours of apprentices and established minimum standards of environmental control in terms of heating, ventilation and humidity control. Enforcement was by means of visitors to factories appointed by local magistrates.

The Factory Act 1833 resulted in the appointment of four factory inspectors with specific powers of entry to factories and means of enforcement. However, it was not until 1864 that such powers were extended beyond the textile industry to the matchmaking and pottery industries and, subsequently, with the Factory and Workshop Act 1878, to other forms of manufacturing industry. A final consolidating Act, the Factory and Workshop Act 1901, formed the basis for much of the current protective health and safety legislation. It enabled the Minister or Secretary of State to make specific regulations covering a wide range of relatively dangerous industrial processes and practices.

Modern health and safety legislation commenced with the Factories Act 1937, the predecessor of the Factories Act 1961. Similar legislation came into operation about this time dealing with mines and quarries, offices, shops and railway premises, and agriculture.

The Report of the Committee on Safety and Health at Work (Cmnd 5034) was published in July 1972 (The Robens Report). This Report proposed substantial changes in the law and administration of occupational health and safety and was instrumental in the passing of the Health and Safety at Work Act 1974, described in the following pages.

Since 1993 European Directives have had a significant effect on UK health and safety legislation.

CRIMINAL AND CIVIL LIABILITY

Breaches of health and safety law can incur both criminal and civil liability.

1. Criminal liability

A crime is an offence against the state. Criminal liability refers to the duties and responsibilities under statute, principally the Health and Safety at Work etc Act 1974, and regulations, and the penalties that can be imposed by the criminal courts, mainly fines and imprisonment.

Criminal law is based on a system of enforcement. Its statutory provisions are enforced by the state's enforcement agencies, such as the police, Health and Safety Executive, local authorities and fire authorities.

2. Civil liability

A civil action generally involves negligence and/or breach of a statutory duty. In such actions a claimant sues a defendant for a remedy (or remedies) that is beneficial to the claimant. In most cases this takes the form of damages, a form of financial compensation. In a substantial number of cases the plaintiff will agree to settle out of court.

Civil liability, therefore, refers to the 'penalty' that can be imposed by a civil court, eg the County Court, High Court, Court of Appeal (Civil Division) or House of Lords, and consists of awards of damages for injury, disease and/or death at work.

HEALTH AND SAFETY AT WORK ACT (HASAWA) 1974

1. Duties of employers

It is the duty of every employer, so far as is reasonably practicable, to ensure the health, safety and welfare at work of all his employees (HASAWA Section 2(1)). In particular, this includes:

(a) the provision and maintenance of plant and systems of work that are safe and without risks to health (HASAWA Section 2(2) (a));
(b) arrangements for ensuring the safety and absence of health risks in connection with the use, handling, storage and transport of articles and substances (HASAWA Section 2(2) (b));
(c) the provision of such information, instruction, training and supervision as is necessary to ensure the health and safety at work of employees (HASAWA Section 2(2) (c));
(d) the maintenance of any place of work under the employer's control in a condition that is safe and without risks to health, and the provision and maintenance of means of access and egress from it that are safe and without risks to health (HASAWA Section 2(2) (d)); and
(e) the provision and maintenance of a working environment for his employees that is safe, without risks to health and adequate as regards facilities and arrangements for their welfare at work (HASAWA Section 2(2) (e)).

2. Duties of employees

It is the duty of every employee while at work:

(a) to take reasonable care for the health and safety of himself and of other persons who may be affected by his acts or omissions at work; and
(b) as regards any duty or requirement imposed on his employer or any other person by or under any of the relevant statutory provisions, to co-operate with him so far as is necessary to enable that duty or requirement to be performed or complied with (HASAWA Section 7).

No person shall intentionally or recklessly interfere with or misuse anything provided in the interests of health, safety or welfare in pursuance of any of the relevant statutory provisions (HASAWA Section 8).

3. Duties of employers to persons other than their employees

Every employer must conduct his undertaking in such a way as to ensure, so far as is reasonably practicable, that persons not in his employment, eg contractors, are not exposed to risks to health or safety (HASAWA Section 3(1)).

4. Duties of occupiers of premises to persons other than their employees

Section 4 of the HASAWA has effect for imposing on persons duties in relation to those who:

(a) are not their employees, but
(b) use non-domestic premises made available to them as a place of work or as a place where they may use plant or substances provided for their use there,

and applies to premises so made available and other non-domestic premises used in connection with them.

In this case the person or persons in control of the premises must take such measures to ensure, so far as is reasonably practicable, that the premises, all means of access thereto or egress therefrom, and any plant or substances in the premises or, as the case may be, provided for use there, is or are safe and without risks to health.

The protection under this section extends to visitors who are:

(a) workers, eg employees of a contract land clearance firm; and
(b) visitors to agricultural premises, eg members of the public invited to pick fruit.

5. Duties of designers, manufacturers, importers and suppliers

Section 6 of the HASAWA originally placed specific duties on the designers, manufacturers, importers and suppliers of articles and substances used at work. These duties were amended by Section 36 and Schedule 3 of the Consumer Protection Act (CPA) 1987 whereby, in the case of *articles* for use at work (see definition below), such persons must:

(a) ensure, so far as is reasonably practicable, that the article is so designed and constructed that it will be safe and without risks to health at all times when it is being set, cleaned, used or maintained by a person at work;
(b) carry out or arrange for the carrying out of such testing and examination as may be necessary for the performance of the duty imposed on him by the preceding paragraph;
(c) take such steps as are necessary to secure that persons supplied by that person with the article are provided with adequate information about the use for which the article is designed or has been tested and about any conditions necessary to ensure that it will be safe and without risks to health at all such times as are mentioned in paragraph (a) above and when it is being dismantled or disposed of; and
(d) take such steps as are necessary to ensure, so far as is reasonably practicable, that persons so supplied are provided with all such revisions of information provided by them by virtue of the preceding paragraph as are necessary by reason of its becoming known that anything gives rise to a serious risk to health or safety.

In the case of *substances* for use at work, such persons have a duty to:

(a) ensure, so far as is reasonably practicable, that the substance will be safe and without risks to health at all times when it is being used, handled, processed, stored or transported by a person at work or in premises to which Section 4 applies;
(b) carry out or arrange for the carrying out of such testing and examination as may be necessary for the performance of the duty imposed on him by the preceding paragraph;
(c) take such steps as are necessary to secure that persons supplied by that person with the substance are provided with adequate information about any risks to health or safety to which the inherent properties of the substance may give rise, about the results of any relevant tests which have been carried out on or in connection with the substance and about any conditions necessary to ensure that the substance will be safe and without risks to health at all such times as are mentioned in paragraph (a) above and when the substance is being disposed of; and
(d) take such steps as are necessary to secure, so far as is reasonably practicable, that persons so supplied are provided with all such revisions of information provided by them by virtue of the preceding paragraph as are necessary by reason of it becoming known that anything gives rise to a serious risk to health or safety.

Under the CPA, 'articles for use at work' and 'substances' are defined thus:
Article for use at work means:

(a) any plant designed for use or operation (whether exclusively or not) by persons at work, or who erect or install any article of fairground equipment; and
(b) any article designed for use as a component in any such plant or equipment.

Substance means any natural or artificial substance (including micro-organisms) intended for use (whether exclusively or not) by persons at work.

SOURCES OF CRIMINAL LAW

1. Statutes

One of the principal functions of Parliament is the enactment of statutes or Acts of Parliament. Most statutes commence their life as a bill and most government-introduced bills start the Parliamentary process in the House of Commons. (Some bills of a non-controversial nature may start in the House of Lords, however.)

The process in the House of Commons commences with a formal first reading. This is followed by a second reading at which stage there is discussion on the general principles and the bill's main purpose. Following the second reading the bill goes to the committee stage for detailed consideration by an appointed committee comprising both Members of Parliament and specialists. Following consideration, the committee reports back to the House with recommendations for amendments. Such amendments are considered by the House and it, in turn, may make amendments at this stage and/or return the bill to the committee for further consideration. After this report stage the bill receives a third reading when only verbal alterations are made.

The bill is then passed to the House of Lords where it goes through a similar process. The House of Lords either passes the bill or amends it, in which case it is returned to the House of Commons for further consideration.

After a bill has been passed by both Houses it receives the Royal Assent, which is always granted, and becomes an Act of Parliament.

2. Regulations

A statute generally confers power on a Minister or Secretary of State to make 'statutory instruments', which may indicate more detailed rules or requirements for implementing the overall objectives of the statute. Most statutory instruments take the form of Regulations and come within the area of law known as delegated or subordinate legislation.

The HASAWA is an enabling Act. This means that, in the case of health and safety regulations, the Secretary of State for Employment has powers conferred under HASAWA (section 15, Schedule 3 and section 80) to make Regulations on a wide variety of issues, but within the general objectives and aims of the parent Act. Examples of Regulations made under HASAWA are the Electricity at Work Regulations 1989, Health and Safety (First Aid) Regulations 1981, and Management of Health and Safety at Work Regulations 1999.

Regulations may:

(a) repeal or modify existing statutory provisions;
(b) exclude or modify the provisions of sections 2–9 of HASAWA or existing provisions in relation to any specified class;
(c) make a specified authority, eg a local authority, responsible for enforcement;
(d) impose approval requirements;
(e) refer to specified documents to operate as references;
(f) give exemptions;
(g) specify the class of person who may be guilty of an offence, eg employers; and
(h) apply to a particular case only.

APPROVED CODES OF PRACTICE

The need to provide elaboration on the implementation of regulations is recognised in Section 16 of the HASAWA, which gives the Health and Safety Commission (HSC) power to prepare and approve codes of practice on matters contained not only in regulations but in Sections 2 to 7 of the Act. Before approving a code, the Health and Safety Executive (HSE), acting for the HSC, must consult with any interested body.

An Approved Code of Practice (ACOP) has a special legal status similar to the Highway Code. No one can be prosecuted for an infringement of an ACOP as such, but if someone is prosecuted for any breach of health and safety law to which an ACOP applies, the ACOP is admissible in evidence, and if the guidance it contains has not been followed, the defendant would need to be able to prove that he had fulfilled the legal requirements in some other way.

Thus an ACOP is a quasi-legal document and, although non-compliance does not constitute a breach, if the contravention of the Act or regulations is alleged, the fact that the ACOP was not followed would be accepted in court as evidence of failure to do all that was *reasonably practicable*. A defence would be to provide that *works of equivalent nature* had been carried out or something equally good or better had been done.

Examples of ACOPs are:

Control of asbestos at work (Control of Asbestos at Work Regulations 2002).
Control of substances hazardous to health (Control of Substances Hazardous to Health Regulations 2002).

HSE Guidance Notes

The HSE issues Guidance Notes in some cases to supplement the information in both regulations and ACOPs. These Guidance Notes have no legal status and are purely of an advisory nature.

HSE Guidance Notes fall into six categories:

1. General safety
2. Chemical safety
3. Environmental hygiene
4. Medical series
5. Plant and machinery
6. Health and safety.

Examples of Guidance Notes are:

EH40	Occupational exposure limits
MS20	Pre-employment health screening
HS(G)37	Introduction to local exhaust ventilation
PM41	Application of photoelectric safety systems to machinery
GS20	Fire precautions in pressurised workings.

POWERS OF INSPECTORS

An inspector appointed under the Act, eg HSE Inspector, Environmental Health Officer, Agricultural Health and Safety Inspector, has the following powers:

(a) to enter premises at any reasonable time, or at any time (day or night) if he has reason to believe a dangerous situation exists;

(b) where an inspector has reason to believe he may be obstructed from entering premises in the exercise of his duties, he may enter the premises accompanied by a police constable; he may also take with him any other person authorised by the enforcing authority, together with any equipment and materials he may need;

(c) to make those examinations and investigations he considers necessary to determine whether there has been a breach of the law;

(d) to direct that premises, or part of a premises, are left undisturbed for so long as is reasonably necessary;

(e) to take measurements and photographs and make such recordings as he considers necessary for the purpose of such investigations or examinations;

(f) to take samples of any articles or substances found in the premises, and of the atmosphere in or in the vicinity of any such premises;

(g) where he has reason to believe that any article or substance has caused or is likely to cause danger to health and safety, to arrange for it to be dismantled or subjected to any process or test;

(h) to take possession of and detain any article or substance for so long as is necessary, either for the purpose of examination or test, to ensure that it is not tampered with before his examination of it is completed, or to ensure that it is available for use as evidence in any subsequent proceedings;

(i) to question any person who may have information relevant to his investigations, either alone or in the presence of another person whom he has invited or allowed to be present, and require that person to answer questions and to sign a declaration of the truth of his answers;

(j) to require the production of, inspect and take copies of or any entry in, any books or documents which are required under the statutory provisions to be kept or which may be necessary for him to see for the purposes of any examination or investigation;

(k) to demand such facilities and assistance as may be necessary to enable him to exercise any of the above-mentioned powers; and

(l) to assume any other powers necessary to enable him to carry into effect the relevant statutory provisions.

If a person in charge of a premises requests to be present at the time, an inspector may not dismantle or subject to any process or test any article or substance other than in the presence of that person. Moreover, where an inspector takes possession of an article or substance, he must leave a notice with a responsible person giving particulars sufficient to identify it.

Where an inspector has reasonable cause to believe that any article or substance found in any premises which he has power to enter is a cause of imminent danger of serious personal injury, he may seize that article or substance and cause it to be rendered harmless. However, where practicable, the inspector must, first of all, take a sample of same and give to a responsible person at the premises a portion of the sample marked in a manner sufficient to identify it, together with a copy of his written report.

In situations where an inspector questions persons mentioned in (i) above, the answers given by that person in the course of the interrogation are not admissible in evidence in any subsequent proceedings against that person.

IMPROVEMENT AND PROHIBITION NOTICES

An inspector appointed under the HASAWA may serve two types of notice.

Improvement notices

Where an inspector is of the opinion that a person:

(a) is contravening one or more of the relevant statutory provisions, or

(b) has contravened one or more of these provisions in circumstances that make it likely that the contravention will continue or be repeated,

he may serve an improvement notice on that person requiring that he remedy the contravention(s) or, as the case may be, the matters occasioning it within such period (ending not earlier than the period within which an appeal may be brought) as may be specified in the notice.

Prohibition notices

Where an inspector is of the opinion that a work activity involves or will involve a risk of serious personal injury, he may serve a prohibition notice on the owner and/or occupier of the premises or the person having control of that activity. Such a notice will direct that the specified activities in the notice shall not be carried on by or under the control of the person on whom the notice is served unless certain specified remedial measures have been complied with.

It should be noted that it is not necessary that an inspector believes that a legal provision is being or has been contravened. A prohibition notice is served where there is an immediate threat to life and in anticipation of danger.

A prohibition notice may have immediate effect after its issue by the inspector. Alternatively, it may be deferred, thereby allowing the person time to remedy the situation, carry out works, etc. The duration of a deferred prohibition notice is stated on the notice.

Failure to comply with notices

In the case of both an improvement notice and a prohibition notice, where there is failure to comply within the time specified, or, in the event of an appeal against the notice, after the expiry of any extra time allowed for compliance by a tribunal, the person concerned can be prosecuted. Where a person is convicted of an offence specified in an improvement or prohibition notice and the contravention is continued after the conviction, he may be found guilty of a further offence and liable on summary conviction to a fine not exceeding £200 for each day on which the contravention is continued, in addition to the original maximum fine of £20,000 for each contravention specified.

Appeals against notices

Where a person lodges an appeal against an improvement notice, the operation of that notice is automatically suspended. In the case of a prohibition notice, however, the requirements of the notice continue to apply, unless a tribunal has directed otherwise, in which case the suspension operates from the date directed by the tribunal.

Prosecution

Prosecution is frequently the outcome of failure to comply with an improvement or prohibition notice. Conversely, an inspector may simply institute legal proceedings without service of a notice. Cases are normally heard in a magistrates court, but there is also provision in the HASAWA on indictment. Much depends upon the gravity of the offence.

HEALTH AND SAFETY EXECUTIVE — Serial No. I

Health and Safety at Work etc. Act 1974, Sections 21, 23 and 24

IMPROVEMENT NOTICE

Name and address (See Section 46)	To ....
	....
(a) Delete as necessary	(a) Trading as ....
(b) Inspector's full name	(b) ....
(c) Inspector's official designation	one of (c) ....
(d) Official address	of (d) ....
	.... Tel No ....
	hereby give you notice That I am of the opinion that ....
(e) Location of premises or place and activity	(e) ....
	you, as (a) an employer/s self employed person/s person wholly or partly in control of the premises
(f) Other specified capacity	(f) ....
	(a) are contravening/have contravened in circumstances that make it likely that the contravention will continue or be repeated
	....
	....
(g) Provisions contravened	(g) ....
	....
	The reasons for my said opinion are:- ....
	....
	....
	and I hereby require you to remedy the said contraventions or, as the case may be, the matters occasioning them by
(h) Date	(h) ....
	(a) In the manner stated in the attached schedule which forms part of the notice.
	Signature .... Date ....
	Being an inspector appointed by an instrument in writing made pursuant to Section 19 of the said Act and entitled to issue this notice.
	(a) An Improvement notice is also being served on
	....
	of ....
LP1	related to the matters contained in this notice.

Figure 1.1 Improvement and prohibition notices

HEALTH AND SAFETY EXECUTIVE Serial No. I

Health and Safety at Work etc. Act 1974, Sections 22–24 Serial No. P

PROHIBITION NOTICE

Name and address (See Section 46)	To ....
	....
(a) Delete as necessary	(a) Trading as ....
(b) Inspector's full name	(b) ....
(c) Inspector's official designation	one of (c) ....
(d) Official address	of (d) ....
	.... tel no ....
	hereby give you notice that I am of the opinion that the following activities
	namely:- ....
	....
	....
	which are (a) being carried on by you/about to be carried on by you/under your control
(e) Location of activity	at (e) ....
	involve, or will involve (a) a risk/an imminent risk, of serious personal injury. I am further of the opinion that the said matters involve contraventions of the following statutory provision:- ....
	....
	....
	....
	because ....
	....
	....
	and I hereby direct that the said activities shall not be carried on by you or under your control (a) immediately/after
(f) Date	(f) ....
	unless the said contraventions and matters included in the schedule, which forms part of this notice, have been remedied.
	Signature .... Date ....
	being an inspector appointed by an instrument in writing made pursuant to Section 19 of the said Act and entitled to issue this notice.
LP2	

Figure 1.1 cont.

Corporate liability

Where an offence under any of the relevant statutory provisions (eg regulations made under HASAWA 1974) committed by a body corporate is proved to have been committed with the consent or connivance of, or to have been attributable to any neglect on the part of, any director, manager, secretary or other similar officer of the body corporate or a person who was purporting to act in any such capacity, he as well as the body corporate shall be guilty of that offence and shall be liable to be proceeded against and punished accordingly. (Sec 37(1) HASAWA).

This means, fundamentally, that:

(a) where an offence is committed through neglect or omission by a constituted board of directors, the company itself can be prosecuted as well as the directors individually who may have been to blame;
(b) where an individual functional director is guilty of an offence, he can be prosecuted as well as the company; and
(c) a company can be prosecuted even though the act or omission resulting in the offence was committed by a junior official or executive, or even a visitor to the company premises.

Other 'corporate' persons, eg personnel managers, training officers, chief engineers, health and safety advisers, may also be guilty of offences. Section 36 of the HASAWA deals with such offences thus:

> Where the commission by any person of an offence under any of the relevant statutory provisions is due to the act or default of some other person, that other person shall be guilty of the offence, and a person may be charged with and convicted of the offence . . . whether or not proceedings are taken against the first mentioned person.

HEALTH AND SAFETY (OFFENCES) ACT 2008

This Act amends section 22 of the HASAWA. It raises the maximum fine that may be imposed by the lower courts from £5,000 to £20,000 for most health and safety offences and makes imprisonment an option for more offences in both the lower and higher courts. Certain offences, which are currently triable only in the lower courts, may be triable in either the lower or higher courts.

The HSE specifies that prosecutions should be in the public interest and where one or more of a list of circumstances apply, namely:

- where death was a result of a breach of the legislation;
- there has been reckless disregard of health and safety requirements;

- there have been repeated breaches which give rise to significant risk, or persistent and significant poor compliance; or
- false information has been supplied willfully, or there has been intent to deceive in relation to a matter which gives rise to significant risk.

CORPORATE MANSLAUGHTER

There have been a number of criminal cases in recent years involving the offence of 'corporate manslaughter'. Fundamentally, manslaughter is of two kinds, that is, voluntary manslaughter and involuntary manslaughter. Voluntary manslaughter, which is essentially murder but reduced in severity owing to, say, diminished responsibility, is not relevant to health and safety. Involuntary manslaughter, on the other hand, extends to all unlawful homicides where there is no malice aforethought or intent to kill.

There are two forms of involuntary manslaughter, that is, constructive manslaughter and reckless manslaughter. Constructive manslaughter applies to situations where death results from an act unlawful at common law or by statute, amounting to more than mere negligence. Reckless manslaughter (or gross negligence) arises where death is caused by a reckless act or omission. A person acts recklessly 'without having given any thought to the possibility of there being any such risk or, having recognised that there was some risk involved, has none the less gone on to take it' (R v Caldwell (1981) 1 AER 961).

Companies, undertakings and their 'controlling minds'

Companies and undertakings operate through their chief executives, directors, managers and employees. In order to convict a company of manslaughter it must be shown that a causal link existed between a grossly negligent act or omission by a person who is the 'controlling mind' of the company and the immediate cause of death.

This principle assumes that there are certain directors or senior managers whose acts and states of mind can properly be regarded as those of the company itself. Such persons are to be identified as those who are entrusted with the powers of the company. Accordingly, the conviction of a company for manslaughter by gross negligence in the absence of evidence establishing the guilt of an identified individual for the same crime is not possible.

Generally, a person can only be a 'controlling mind' if he is a director or other superior officer, undertaking the functions of management, but if the board has delegated part of its functions of management to a named individual, with full discretion to act independently, then in such circumstances, the delegated person can be considered a controlling mind. (Tesco v Nattrass (1971) 2 AER).

In these circumstances the prosecution will need to show that an act or omission by a controlling officer created a dangerous situation and whether precautions were then taken to guard against that danger. With large organisations featuring a diffuse management structure, it is more difficult to attribute grossly negligent acts or omissions committed in the course of the company's operations to a controlling officer and, therefore, to the organisation itself. Conversely, with a small company, it would be relatively straightforward to establish the causal link.

CORPORATE MANSLAUGHTER AND CORPORATE HOMICIDE ACT 2007

The offence

This statute sets out a specific offence for convicting an organisation where a gross failure in the way activities were managed or organised resulted in:

- a person's death; and
- amounts to a gross breach of a relevant duty of care owed by the organisation to the deceased.

This offence is called 'corporate manslaughter' in England, Wales and Northern Ireland, and 'corporate homicide' in Scotland.

Organisations

The Act applies to:

- companies incorporated under companies legislation or overseas;
- other corporations including:
  a. public bodies incorporated by statute such as local authorities, NHS bodies and a wide range of non-departmental public bodies;
  b. organisations incorporated by Royal Charter;
  c. limited liability partnerships.
- all other partnerships and trade unions and employer's associations, if the organisation concerned is an employer;
- Crown bodies, such as government departments or police forces.

Management systems

Thus, an organisation to which this offence applies is guilty of an offence if the way its activities are managed or organised by its senior management is a substantial element in the breach of the law stated above. It is committed by the persons who make significant decisions about the organisation or substantial parts of it. This includes both centralised headquarters functions as well as those in operational management roles.

Fundamentally, the organisation's conduct must have fallen far below that which could have been reasonably expected. Judges must take into account any breaches of health and safety legislation by the organisation and how serious and dangerous those failures were.

Relevant duty of care

A relevant duty of care in relation to an organisation means any of the following duties owed by it under the law of negligence:

(a) a duty owed to its employees or to other persons working for the organisation or performing services for it;
(b) a duty owed as occupier of premises;
(c) a duty owed in connection with:
    (i) the supply by the organisation of goods or services (whether for consideration or not);
    (ii) the carrying on by the organisation of any construction or maintenance operations;
    (iii) the carrying on by the organisation of any other activity on a commercial basis; or
    (iv) the use or keeping by the organisation of any plant, vehicle or other thing.
(d) a duty owed by a person who, by reason of being a person within ss(2) is someone for whose safety the organisation is responsible.

Gross breach of a duty of care

Section 8 of the Act outlines the factors a jury must consider where an organisation owed a relevant duty of care to a person and it falls to the jury to decide whether there has been a gross breach of that duty.

The jury must consider whether the evidence shows that the organisation failed to comply with any health and safety legislation that relates to the alleged breach, and if so:

(a) how serious that failure was;
(b) how much of a risk of death it posed.

The jury may also:
(a) consider the extent to which the evidence shows that there were attitudes, policies, systems or accepted practices within the organisation that were likely to have encouraged any such failure as is mentioned above, or to have produced tolerance of it;

(b) have regard to any health and safety guidance that relates to the alleged breach.

Health and safety guidance means any code, guidance, manual or similar publication that is concerned with health and safety matters and is made or issued (under a statutory provision or otherwise) by an authority responsible for the enforcement of any health and safety legislation.

Remedial Orders

A court before which an organisation is convicted of corporate manslaughter or corporate homicide may make a Remedial Order requiring the organisation to take specified steps to remedy:

(a) the relevant breach;
(b) any matter that appears to the court to have resulted from the relevant breach and to have been a cause of death;
(c) any deficiency, as regards health and safety matters, in the organisation's policies, systems or practices of which the relevant breach appears to the court to be an indication.

Publicity Orders

A court before which an organisation is convicted of corporate manslaughter or corporate homicide may make a Publicity Order requiring the organisation to publicise in a specified manner:

(a) the fact that it has been convicted of the offence;
(b) specified particular of the offence;
(c) the amount of any fine imposed;
(d) the terms of any Remedial Order made.

Individual liability

An individual cannot be guilty of aiding, abetting, counselling or procuring the commission of the offence of corporate manslaughter or corporate homicide.

Penalties

An organisation convicted of this offence can receive:

(a) a fine for which no upper limit is prescribed;
(b) a Publicity Order; and
(c) a Remedial Order.

Enforcement

Investigations relating to corporate manslaughter/homicide are led by the Police and, in certain cases, with the assistance of the HSE, local authorities and other agencies.

LEADERSHIP ACTIONS FOR DIRECTORS AND BOARD MEMBERS

The combined Institute of Directors/HSE publication *Leading Health and Safety at Work* (2009) sets out an agenda for the effective leadership of health and safety. This Guidance is designed for use by all directors, governors, trustees, officers and their equivalents in the private, public and third sectors. It applies to organisations of all sizes.

Essential principles

The Guidance establishes three essential principles:

1. Strong and active leadership from the top:
   - visible, active commitment from the board;
   - establishing effective 'downward' communication systems and management structures;
   - integration of good health and safety management with business decisions.
2. Worker involvement
   - engaging the workforce in the promotion and achievement of safe and healthy conditions;
   - effective 'upward' communication;
   - providing high quality training.
3. Assessment and review
   - identifying and managing health and safety risks;
   - accessing (and following) competent advice;
   - monitoring, reporting and reviewing performance.

Health and safety leadership checklist

The Guidance finishes with a list designed to check an employer's status as a leader on health and safety.

1. How do you demonstrate the board's commitment to health and safety?
2. What do you do to ensure appropriate board-level review of health and safety?

3. What have you done to ensure your organisation, at all levels including the board, receives competent health and safety advice?
4. How are you ensuring all staff, including the board, are sufficiently trained and competent in their health and safety responsibilities?
5. How confident are you that your workforce, particularly safety representatives, are consulted properly on health and safety matters, and that their concerns are reaching the appropriate level including, as necessary, the board?
6. What systems are in place to ensure your organisation's risks are assessed, and that sensible control measures are established and maintained?
7. How well do you know what is happening on the ground, and what audits or assessments are undertaken to inform you about what your organisation and contractors actually do?
8. What information does the board receive regularly about health and safety, eg performance data and reports on injuries and work-related ill health?
9. What targets have you set to improve health and safety and do you benchmark your performance against others in your sector or beyond?
10. Where changes in working arrangements have significant implications for health and safety, how are these brought to the attention of the board?

JOINT CONSULTATION WITH EMPLOYEES

The Safety Representatives and Safety Committees Regulations (SRSCR) 1977 provide for the appointment in prescribed cases by recognised trade unions of safety representatives from among the employees. Those selected represent the workforce in consultation with the employer with a view to promoting and developing measures to ensure the health and safety at work of employees, and in checking the effectiveness of such measures.

1. Safety representatives

A safety representative is a person appointed by his trade union to represent workers in consultations with the employer on all matters relating to health and safety at work. He has the following specific functions, following notification to the employer in writing of his appointment:

(a) to investigate potential hazards and causes of accidents at the workplace;
(b) to investigate complaints from employees concerning health and safety risks at work;
(c) to make representation to the employer on matters arising out of (a) and (b) above and on general matters affecting the health, safety and welfare of employees;

(d) to carry out certain inspections:
   - of the workplace, after giving reasonable notice to the employer;
   - of a relevant area following a reportable accident or scheduled dangerous occurrence, or where a reportable disease is contracted, if it is safe to do so and in the interests of the employees represented;
   - of documents relating to the workplace or employees which the employer is required to maintain;

(e) to represent his group of employees in consultation with inspectors appointed under the Act, and to receive information from them;

(f) to attend meetings of safety committees.

2. Safety committees

Where requested in writing by at least two safety representatives the employer must form a safety committee (SRSCR). When establishing a safety committee, the employer must:

(a) consult with both the safety representatives making this request, and with representatives of the trade union whose members work in any workplace where it is proposed that the committee will function;

(b) post a notice stating the composition of the committee and the workplace(s) to be covered by it, in a place where it can easily be read by employees;

(c) establish the committee within three months following the request for its formation.

It is the employer's prerogative to establish the objectives, role and function of the safety committee, together with aspects such as frequency of meeting, rotation of chairman, minutes, agenda, etc.

It should be appreciated that safety committees have a significant role in reducing accidents and occupational ill health. They are a forum for discussion on a wide range of topics, an important system for dissemination of information and can be extremely effective in maintaining good standards of communication between employers and employees.

3. Non-unionised employees

Under the Health and Safety (Consultation with Employees) Regulations 1996, employers must consult any employees who are not covered by the Safety Representatives and Safety Committees Regulations. This may be by direct consultation with employees or through representatives elected by the employees they are to represent.

HSE Guidance

HSE Guidance accompanying the regulations details:

(a) which employees must be involved;
(b) the information they must be provided with;
(c) procedures for the election of representatives of employee safety;
(d) the training, time off and facilities they must be provided with; and
(e) their functions in office.

RUNNING A SAFETY COMMITTEE

As with any committee, it is essential that the constitution of a safety committee be in written form. The following points should be considered when establishing and running a safety committee.

1. Objectives

To monitor and review the general working arrangements for health and safety.

To act as a focus for joint consultation between employer and employees in the prevention of accidents, incidents and occupational ill health.

2. Composition

The composition of the committee should be determined by local management, but should normally include equal representation of management and employees, ensuring all functional groups are represented.

Other persons may be co-opted to attend specific meetings, eg health and safety adviser, company engineer.

3. Election of committee members

The following officers should be elected for a period of one year: the chairman, the deputy chairman and the secretary. Nominations for these posts should be submitted by a committee member to the secretary for inclusion in the agenda of the final meeting in each yearly period. Members elected to office may be renominated or re-elected to serve for further terms.

Election should be by ballot and should take place at the last meeting in each yearly period.

4. Frequency of meetings

Meetings should be held on a quarterly basis or according to local needs. In exceptional circumstances extraordinary meetings may be held by agreement of the chairman.

5. Agenda and minutes

The agenda should be circulated to all members at least one week before each committee meeting and should include the following items:

(a) Apologies for absence

Members unable to attend a meeting should notify the secretary and make arrangements for a deputy to attend on their behalf.

(b) Minutes of the previous meeting

Minutes of the meeting should be circulated as widely as possible and without delay. All members of the committee, senior managers, supervisors and trade union representatives should receive personal copies. Additional copies should be posted on notice boards.

(c) Matters arising

The minutes of each meeting should incorporate an 'action column' in which persons identified as having future action to take, as a result of the committee's decisions, are named.

The named person should submit a written report to the secretary, which should be read out at the meeting and included in the minutes.

(d) New items

Items for inclusion on the agenda should be submitted to the secretary, in writing, at least seven days before the meeting. The person requesting the item for inclusion on the agenda should state in writing what action has already been taken through the normal channels of communication. The chairman will not normally accept items that have not been pursued through the normal channels of communication prior to submission to the secretary.

Items requested for inclusion after the publication of the agenda should be dealt with, at the discretion of the chairman, as emergency items.

(e) Safety adviser's report

The safety adviser should submit a written report to the committee, copies of which should be issued to each member at least two days prior to the meeting and attached to the minutes. The safety adviser's report should include, for example:

(i) a description of all reportable injuries, diseases and dangerous incidents that have occurred since the last meeting, together with details of remedial action taken;
(ii) details of any new health and safety legislation directly or indirectly affecting the organisation, together with details of any action that may be necessary;
(iii) information on the outcome of any safety monitoring activities undertaken during the month, eg safety inspections of specific areas, risk assessments undertaken; and
(iv) any other matters that, in the opinion of the secretary and himself, need a decision from the committee.

(f) Date, time and place of the next meeting

STATEMENTS OF HEALTH AND SAFETY POLICY

Section 2(3) of the HASAWA imposes a duty on every employer of five or more persons to prepare, and bring to the notice of his employees, a written statement of his general policy with respect to the health and safety at work of his employees. Guidance is provided in Leaflet No HSC 6 *Guidance Notes on Employers' Policy Statements for Health and Safety at Work* and *Writing Your Health and Safety Policy Statement – How to prepare a safety policy statement for a small business*, both available from HMSO.

Content of the policy statement

Essentially, a statement of health and safety policy should be in written form and signed and dated by the owner, occupier or person having control of a business. It should consist of three parts principally, and be subject to regular revision according to individual circumstances, eg change of individual responsibilities. It is standard practice, further, to incorporate a number of appendices dealing with specific aspects of the policy statement, in particular the individual responsibilities of all levels of management and workers for health and safety. The three principal parts of a statement of health and safety policy are as follows.

1. A general statement of intent

This should outline in broad terms the company's overall philosophy in relation to the management of health and safety. It should also include broad reference to the responsibilities of directors, line management and employees.

2. Organisation

This part is concerned with people and their duties, and outlines the chain of command in terms of health and safety management, in particular individual accountabilities, the system for monitoring implementation of the policy, the role and function of trade union safety representatives and of the safety committee, and a management chart showing the lines of responsibility from managing director or chief executive downwards.

3. Arrangements

This part deals with the systems and procedures for ensuring appropriate standards of safety, health and welfare, including the practical arrangements for their implementation.

Aspects to be incorporated in the 'Arrangements' include, for instance, the system for health and safety training, control of the working environment, procedures for operating safe systems of work, machine guarding, housekeeping procedures, noise control, dust control, fire protection procedures, health surveillance arrangements, systems for reporting, recording and investigation of accidents, ill health and dangerous occurrences, emergency procedures, eg in the event of fire, and safety monitoring systems. This part of the policy statement, above all, must be a 'living document' which is subject to regular revision and updating.

Fundamentally, a statement of health and safety policy is specific to a particular premises or business operation. It must be carefully thought out, there must be consultation with the workforce during its preparation if it is to be successful and, above all, everyone must be fully aware of his responsibilities towards himself and others with the principal aim of preventing accidents, ill health and other loss-producing incidents.

Appendices to the policy statement

Where specific attention to a particular aspect is necessary, it is common practice to add a number of appendices to the policy statement. Such appendices could include:

(a) general and specific legislation applying to the business;
(b) individual responsibilities of management and employees;
(c) specific company policies relating to, for instance, Aids, smoking on work premises, pre-employment health examinations and other forms of health surveillance, the provision and use of certain forms of personal protective equipment, eg eye protection, first aid procedures;

(d) the hazards that could be encountered and the precautions necessary on the part of workers, eg from machinery, dangerous substances, or the use of ladders;
(e) joint consultation procedures;
(f) the system for providing health and safety information to employees;
(g) fatal and major injury accident procedures;
(h) procedures to protect visitors, eg contractors, and members of the public from risks arising from work activities.

The hierarchy of duties

The duties of people at work under the FA, HASAWA, etc, vary. Certain legal duties on employers and others may be of an absolute nature, or qualified by the terms 'where practicable' or 'so far as is reasonably practicable'.

'Absolute' or 'strict' duties: where there is a high degree of risk of death or serious injury if safety precautions are not taken, as in the case of certain classes of machinery, the duty on an employer may well be of an absolute or strict nature. Such duties are laid down, for example, in the Provision and Use of Work Equipment Regulations 1998, which places an absolute duty on an employer to 'ensure that work equipment is maintained in an efficient state, in efficient working order and in good repair'.

'Practicable' means more than physically possible. Thus in Adsett v Steel Founders Ltd, the judge said that the measures must be able to be carried out 'in the light of current knowledge and invention'. Moreover, it was pointed out that 'practicable' implies a higher standard of care than the term 'reasonably practicable'.

'Reasonably practicable' means a comparison must be made between, on the one hand, the extent of the risk and, on the other, the sacrifice (costs, time and effort) in taking the measures necessary to avert the risk. If it can be shown that there is a disparity between the degree of risk and sacrifice involved, then the defendant has discharged the burden of proof.

Both the above expressions are concerned with the burden of proof, the former imposing a higher standard than the latter.

'All reasonable precautions and all due diligence'

An important defence, taken from food safety legislation, has been incorporated in recent health and safety legislation, such as the Control of Substances Hazardous to Health (COSHH) Regulations 2002, Electricity at Work Regulations 1989 and the Pressure Systems Safety Regulations 2000. This defence makes provision for a person charged with an offence to plead that he took '**all** reasonable precautions or steps and exercised all due diligence' to prevent the commission of the offence. The significance of the word 'all' in each case should be noted, and if such a defence is to be successful, proof of management and operation of many of the following systems and procedures, and supported by appropriate documentation, eg internal codes of practice, is necessary:

1. A well-written statement of health and safety policy, which clearly identifies individual responsibilities of all concerned.
2. Preventive maintenance schedules.
3. Cleaning schedules.
4. Procedures for providing information, instruction and training to staff, contractors and visitors, in particular, induction and orientation training procedures.
5. Formally documented safe systems of work, together with Permit to Work systems.
6. Identification of persons classed as 'competent persons' together with their specific responsibilities.
7. Safety monitoring procedures, eg safety audits.
8. Accident and incident reporting, recording and investigation procedures.
9. Occupational health and hygiene procedures.
10. Environmental control systems, eg noise control.
11. Procedures for vetting the relative safety of new machinery, plant, equipment and potentially dangerous substances.
12. Procedures for regulating the activities of contractors.
13. Fire protection and evacuation procedures.
14. Procedures for liaison with enforcement agencies, eg Health and Safety Executive, and action to be taken following the service of an Improvement or Prohibition Notice.
15. Health and safety promotional activities, eg health and safety awards.

Documentation of the above is extremely important, and many of the above policies, systems and procedures should be incorporated in an organisation's Health and Safety Manual which should be related to and referred to in the Statement of Health and Safety Policy.

MANAGEMENT OF HEALTH AND SAFETY AT WORK REGULATIONS 1999

These regulations are accompanied by an Approved Code of Practice (ACOP) produced by the Health and Safety Commission (HSC). They are of particular significance in view of the fact that all post-1992 regulations must be read in conjunction with the duties laid down under these, with particular reference to risk assessment.

Duties on individuals are of an absolute or strict nature, compared with those under the HASAWA and other regulations where the duties are qualified by the term 'so far as is practicable' or 'so far as is reasonably practicable'.

The principal features of the regulations are outlined below.

Risk assessment

Every employer shall make a suitable and sufficient assessment of:

(a) the risks to the health and safety of his employees to which they are exposed while at work; and
(b) the risks to the health and safety of persons not in his employment arising out of or in connection with the conduct by him of his undertaking,

for the purpose of identifying the measures he needs to take to comply with the requirements and prohibitions imposed upon him by or under the *relevant statutory provisions*.

These are the provisions of the statute, ie the HASAWA, and of any regulations made under the statute, eg Control of Noise at Work Regulations 2005. Similar provisions apply in the case of self-employed persons. (Regulation 3(2))

Any assessment shall be reviewed by the employer if there is reason to suspect it is no longer valid or there has been a significant change in the matters to which it relates. (Regulation 3(3))

An employer shall not employ a young person, ie between the ages of 16 and 18 years, unless he has, in relation to the risks to the health and safety of young persons, made or reviewed an assessment. (Regulation 3(4))

In making or reviewing the assessment an employer who employs or is to employ a young person shall take particular account of:

(a) the inexperience, lack of awareness of risks and immaturity of young persons;
(b) the fitting-out and layout of the workplace and the workstation;
(c) the nature, degree and duration of exposure to physical, chemical and biological agents;
(d) the form, range and use of work equipment and the way in which it is handled;
(e) the organisation of processes and activities;
(f) the extent of the health and safety training provided or to be provided to young persons; and

(g) risks from agents, processes and work listed in the Annex to Council Directive 94/33/EC on the protection of young people at work. (Regulation 3(5))

Principles of prevention to be applied

Where an employer implements preventive and protective measures he shall do so on the basis of the principles specified in Schedule 1 to the Regulations. (Regulation 4)

Health and safety arrangements

Every employer shall make and give effect to such arrangements as are appropriate, having regard to the nature of his activities and the size of his undertaking, for the *effective planning, organisation, control, monitoring and review* of the preventive and protective measures. These arrangements must be recorded where five or more employees are employed. (Regulation 5)

Health surveillance

Every employer shall ensure that his employees are provided with such health surveillance as is appropriate, having regard to the risks to their health and safety which are identified by the assessment. (Regulation 6)

Health and safety assistance

Every employer must appoint one or more competent persons to assist him in undertaking the measures he needs to take to comply with the requirements and prohibitions imposed upon him by or under the relevant statutory provisions.

Where an employer appoints competent persons, he shall make arrangements for ensuring adequate co-operation between them. (Regulation 7(2))

The employer shall ensure that the number of persons appointed, the time available for them to fulfil their functions and the means at their disposal are adequate having regard to the size of his undertaking, the risks to which his employees are exposed and the distribution of those risks throughout the undertaking. (Regulation (3))

A person shall be regarded as competent where he has sufficient training and experience or knowledge and other qualities to enable him properly to assist in undertaking the measures referred to above. (Regulation 7(5))

Procedures for serious and imminent danger and for danger areas

Employers must:

(a) establish and where necessary give effect to appropriate procedures to be followed in the event of serious or imminent danger to persons at work;
(b) nominate a sufficient number of *competent persons* to implement these procedures;
(c) prevent any employee being given access to a danger area unless he has received adequate health and safety instruction. (Regulation 8(1))

Contacts with external services

Every employer shall ensure that any necessary contacts with external services are arranged, particularly as regards first aid, emergency medical care and rescue work. (Regulation 9)

Information for employees

Every employer shall provide his employees with *comprehensible* and *relevant* information on:

(a) the risks to their health and safety identified by the assessment;
(b) the preventive and protective measures;
(c) the procedures referred to in Regulation 8(1)(a);
(d) the identity of the competent persons nominated in accordance with Regulation 8(1)(b); and
(e) the risks notified to him in accordance with Regulation 11(1)(c) (shared workplaces). (Regulation 10(1))

Every employer shall, before employing a child, provide a parent of the child with comprehensible and relevant information on:

(a) the risks to his health and safety identified by the assessment;
(b) the preventive and protective measures; and
(c) the risks notified in accordance with Regulation 11(1)(c). (Regulation 10(2))

Co-operation and co-ordination

This regulation concerns the duties of employers who share workplaces, whether on a temporary basis, eg a construction site, or a permanent basis, eg office block, industrial estate. Each employer must:

(a) co-operate with other employers to enable them to comply with legal requirements;
(b) take all reasonable steps to co-ordinate the measures he is taking with other employers in order to comply with legal requirements; and
(c) take all reasonable steps to inform other employers of risks to their employees' health and safety arising from his activities. (Regulation 11(1))

Note: These duties should be read in conjunction with the requirements under the Construction (Design and Management) Regulations 2007 where a construction project may involve several employers.

Persons working in host employers or self-employed persons' undertakings

Every employer and self-employed person shall ensure that the employer of any employees from an outside undertaking who are working in his undertaking is provided with comprehensible information on:

(a) the risks arising in the undertaking; and
(b) the measures taken by the first-mentioned employer to comply with legal requirements and to protect those employees.

These measures include details of emergency and evacuation procedures and the competent persons nominated to implement such procedures. (Regulation 12)

Capabilities and training

Every employer shall, in entrusting tasks to his employees, take into account their capabilities as regards health and safety. (Regulation 13(1))

Note: This requirement implies a need to match the individual to the task, both mentally and physically, from a health and safety viewpoint. It requires a good understanding of the human factors aspect of health and safety, in particular ergonomic considerations, the physical and mental limitations of people and the potential for human error.

Every employer shall ensure that his employees are provided with adequate health and safety training:

(a) on their being recruited into the employer's undertaking;
(b) on their being exposed to new or increased risks because of:
    (i) their being transferred or given a change of responsibilities within the employer's undertaking;
    (ii) the introduction of new work equipment into or a change respecting work equipment already in use within the employer's undertaking;

(iii) the introduction of new technology into the employer's undertaking; or
(iv) the introduction of a new system of work or a change respecting a system of work already in use within the employer's undertaking. (Regulation 13(2))

Training shall:

(a) be repeated periodically where appropriate;
(b) be adapted to take account of new or changed risks; and
(c) take place during working hours. (Regulation 13(3))

Employees' duties

Every employee shall use any machinery, equipment, dangerous substance, transport equipment, means of production or safety device provided to him by his employer in accordance both with any training in the use of the equipment concerned which has been received by him and the instructions respecting that use which have been provided to him by the said employer in compliance with the requirements and prohibitions imposed upon that employer by or under the relevant statutory provisions. (Regulation 14(1))

Every employee shall inform his employer or any other employee with specific responsibility for the health and safety of his fellow employees (eg competent person, trade union safety representative):

(a) of any work situation which a person with the first-mentioned employee's training and instruction would reasonably consider represented a serious and immediate danger to health and safety; and
(b) of any matter which a person with the first-mentioned employee's training and instruction would reasonably consider represented a shortcoming in the employer's protection arrangements for health and safety. (Regulation 14(2))

Temporary workers

Every employer shall provide any person working under a fixed-term contract of employment or employed in an employment business with comprehensible information on:

(a) any special occupational qualifications or skills required to be held by that employee if he is to carry out his work safely; and
(b) of any health surveillance required,

before the employee concerned commences his duties. (Regulation 15(1))

Risk assessment in respect of new or expectant mothers

Where:

(a) the persons working in an undertaking include women of child-bearing age; and
(b) the work is of a kind which could involve risk, by reason of her condition, to the health and safety of a new or expectant mother, or to that of her baby, from any processes or working conditions, or physical, biological or chemical agents,

the assessment required by Regulation 3(1) shall also include an assessment of such risk. (Regulation 16(1))

Where, in the case of an individual employee, the taking of any other action the employer is required to take under the relevant statutory provisions would not avoid the risk referred to above, the employer shall, if it is reasonable to do so, and would avoid such risks, alter her working conditions or hours of work. (Regulation 16(2))

If it is not reasonable to alter working conditions or hours of work, or if it would not avoid such risk, the employer shall, subject to section 67 of the Employment Rights Act (ERA) 1996, suspend the employee from work so long as is necessary to avoid such risk. (Regulation 16(3))

References to risk, in relation to risk from any infectious or contagious disease, are references to a level of risk which is in addition to the level to which a new or expectant mother may be expected to be exposed outside the workplace. (Regulation 16(4))

Certificate from a registered medical practitioner in respect of new or expectant mothers

Where:

(a) a new or expectant mother works at night; and
(b) a certificate from a registered medical practitioner or a registered midwife shows that it is necessary for her health or safety that she should not be at work for any period of such work identified in the certificate,

the employer shall, subject to section 67 of the ERA 1996, suspend her from work so long as is necessary for her health or safety. (Regulation 17(1))

Notification by new or expectant mothers

Nothing in paragraph 2 or 3 of Regulation 16 shall require the employer to take any action in relation to an employee until she has notified the employer in writing that she is pregnant, has given birth within the previous six months, or is breastfeeding. (Regulation 18(1))

Nothing in paragraph 2 or 3 of Regulation 16 or in Regulation 17 shall require the employer to maintain action taken in relation to an employee:

(a) in a case:
  (i) to which Regulation 16(2) or (3) relates; and
  (ii) where the employee has notified her employer that she is pregnant, where she has failed, within a reasonable time of being requested to do so in writing by her employer, to produce for the employer's inspection a certificate from a registered medical practitioner or a registered midwife showing that she is pregnant;
(b) once the employer knows that she is no longer a new or expectant mother; or
(c) if the employer cannot establish whether she remains a new or expectant mother. (Regulation 18(2))

Protection of young persons

Every employer shall ensure that young persons employed by him are protected at work from any risks to their health or safety which are a consequence of their lack of experience, or absence of awareness of existing or potential risks or the fact that young persons have not yet fully matured. (Regulation 19(1))

Subject to paragraph 3, no employer shall employ a young person for work:

(a) which is beyond his physical or psychological capacity;
(b) involving harmful exposure to agents which are toxic or carcinogenic, cause heritable genetic damage or harm to the unborn child or which in any other way chronically affect human health;
(c) involving harmful exposure to radiation;
(d) involving the risk of accidents which it may reasonably be assumed cannot be recognised or avoided by young persons owing to their insufficient attention to safety or lack of experience or training; or
(e) in which there is a risk to health from:
  (i) extreme cold or heat,
  (ii) noise, or
  (iii) vibration,

and in determining whether work will involve harm or risks for the purpose of this paragraph, regard shall be had to the results of the assessment. (Regulation 19(2))

Nothing in paragraph 2 shall prevent the employment of a young person who is no longer a child for work:

(a) where it is necessary for his training;
(b) where the young person will be supervised by a competent person; and
(c) where any risk will be reduced to the lowest level that is reasonably practicable. (Regulation 19(3))

Provisions as to liability

Nothing in the relevant statutory provisions shall operate so as to afford an employer a defence in criminal proceedings for a contravention of those provisions by reason of any act or default of:

(a) an employee of his; or
(b) a person appointed by him under Regulation 7 (competent person). (Regulation 21)

Restriction of civil liability for breach of statutory duty

Breach of a duty imposed on an employer by these regulations shall not confer a right of action in any civil proceedings in so far as that duty applies for the protection of a third party.

Breach of a duty imposed on any employee by Regulation 14 (employees' duties) shall not confer a right of action in any civil proceedings in so far as that duty applies for the protection of a third party.

In this regulation, third party, in relation to the undertaking, means any person who may be affected by that undertaking other than the employer whose undertaking it is and persons in his employment.

SCHEDULE 1

General principles of prevention

Schedule 1 specifies the general principles of prevention set out in Article 6(2) of Council Directive 89/391/EEC thus:

(a) avoiding risks;
(b) evaluating the risks that cannot be avoided;
(c) combating the risks at source;
(d) adapting the work to the individual, especially as regards the design of workplaces, the choice of work equipment and the choice of working and production methods, with a view, in particular, to alleviating monotonous work and work at a pre-determined work rate and reducing their effect on health;
(e) adapting to technical progress;
(f) replacing the dangerous by the non-dangerous or the less dangerous;

(g) developing a coherent overall prevention policy that covers technology, organisation of work, working conditions, social relationships and the influence of factors relating to the working environment;
(h) giving collective protective measures priority over individual protective measures; and
(i) giving appropriate instructions to employees.

Work away from base

A substantial number of people work away from their normal base of operations, such as building contractors, people involved in the installation and servicing of equipment, drivers and company representatives. As such, they are exposed to a wide range of hazards through, in most cases, their unfamiliarity with premises, processes and working practices. In some organisations around 20 to 25 per cent of accidents to staff take place on other people's premises.

The legal requirements relating to work on other people's premises are dealt with in the Occupiers' Liability Act 1957, HASAWA and the MHSWR.

Occupiers' Liability Act 1957

An occupier of premises owes a *common duty of care* to all *lawful* visitors in respect of dangers due to the *state* of the premises or *things done or omitted to be done* on them.

Health and Safety at Work etc Act 1974 (HASAWA)

An employer must conduct his undertaking in such a way as to ensure, so far as reasonably practicable, that *persons not in his employment* who may be affected thereby are not thereby exposed to risks to their health or safety. (Section 3)

Every person who has, to any extent, control of premises must ensure, so far as is reasonably practicable, that the premises, all means of access thereto and egress therefrom, and any plant or substances in the premises or provided for use there, is or are safe and without risks to health. (Section 4)

Management of Health and Safety at Work Regulations 1999

Regulation 12 deals with *persons working in host employers' undertakings*. Host employers must provide such persons with *appropriate instructions and comprehensible information* on the risks arising out of or in connection with his undertaking and, secondly, the measures taken in compliance with the requirements and prohibitions imposed upon him by or under the relevant statutory provisions insofar as the said requirements and prohibitions relate to those employees.

Summary

The duties of employers and occupiers of premises towards non-employees working in their premises or undertaking can be summarised as follows:

(a) a general duty of care to all lawful visitors;
(b) a general duty not to expose non-employees to risks to their health or safety;
(c) a general duty on controllers of premises to provide safe premises, safe access and egress, safe plant and substances;
(d) a specific duty to provide instructions and information on risks and precautionary measures necessary by the non-employees or self-employed persons concerned.

Practical procedures to implement these requirements

1. Provision of written instructions to non-employees with regard to safe working practices generally.
2. Provision of comprehensible written information to non-employees on the hazards and precautions necessary.
3. Formal health and safety training sessions for all non-employees prior to commencing work on the host employer's premises.
4. Specification of the health and safety competence necessary for non-employees at the tender stage of contracts.
5. Operation of formal hazard reporting systems by non-employees.
6. Disciplinary procedures against non-employees for a failure to comply with written instructions and safety signs, including dismissal from the site or premises in serious cases of non-compliance.
7. General supervision of non-employees and regular meetings to reinforce the safety requirements for such persons.
8. Liaison with the external employers, and certification where necessary, to ensure the employees concerned have received the appropriate information, instruction and training necessary prior to commencing work in the host employer's undertaking.
9. Pre-tender and ongoing site inspections by the external employer to ensure his employees are not exposed to risks to their health or safety.

PASSPORT SCHEMES

Many employees work away from their main workplace, such as those involved in contracting activities involving the servicing of plant and equipment, construction, stripping of asbestos, contract catering and cleaning activities. Apart from the general duty on employers under the HASAWA for such employees to be provided with health and safety training, many regulations, such as the COSHH Regulations, lay down specific requirements for the training of employees.

So how does a client, in selecting a competent contractor for construction work at his premises, or prior to taking on a contract catering service, ensure that the employees of that contractor are adequately trained in health and safety procedures and the precautions necessary to ensure safe working? One of the ways is through the operation of a safety 'passport' scheme.

Passport schemes ensure that workers have received health and safety awareness training, and are particularly useful for workers who work in more than one industry or organisation. Passport schemes operate in a number of ways. In the majority of cases such schemes are driven by a particular industry, based on the need to ensure that the employees in that industry, suppliers of services, contractors, self-employed persons and agency staff meet a particular training standard. On this basis, the industry may decide:

- what training is required, particularly core syllabus requirements;
- the qualifications and resources needed by trainers;
- how training will be delivered and assessed, perhaps through passing an accredited training course;
- for how long a passport will be valid;
- the need for refresher training before renewal; and
- the system for keeping records.

Following the development of the training course, courses are offered to workers, who must pass some form of assessment before a passport is issued. On satisfactory completion of the course, the worker is issued with a card, similar to a photo-card driving licence, bearing his photograph, signature and a date of expiry of the card.

After the passport scheme has come into operation, only those holding a valid passport are allowed access to workplaces or construction sites. In some cases, clients insist that all contractors' employees hold Passports before completing the selection process for a principal contractor.

Outcome of a passport training scheme

On completion of training, passport holders should know:

- the hazards and risks they may face;
- the hazards and risks they can cause for other people;
- how to identify relevant hazards and potential risks;
- how to assess what to do to eliminate the hazards and control the risk;
- how to take steps to control the risks to themselves and others;
- their safety and environmental responsibilities, and those of the people they work with;

- where to find extra information they need to do their job safely; and
- how to follow a safe system of work.

Co-operation between organisations

The HSE encourages organisations to co-operate in this scheme, whereby one scheme recognises the core training of other schemes. This means that passport holders do not have to repeat the core syllabus if they move from one employer or contract to another. They will simply need site and/or activity-specific training.

Monitoring

It is important that supervisors monitor passport holders on a day-to-day basis by asking people about their work, checking whether people are following procedures and observing their work. The standard of training should further be monitored by clients.

For further information see *Passport Schemes for Health, Safety and the Environment: A good practice guide*, INDG381, HSE Books.

HOME WORKING

Many employees work at home with little or no direct contact with their employer. The scale of work activities undertaken by home workers is extensive, such as the packaging of items such as screws for national store chains; assembly, finishing and packing of electrical goods; work with computers; telephone contact work, such as tele-sales operations; and other tasks, such as ironing and repairing clothing. In the case of those involved in assembly work, components may be delivered, and finished items collected, on a weekly basis. Other home workers may receive information on their future work requirements by letter or e-mail.

The hazards arising from home working activities

The hazards to home workers are extensive. These may include electrical hazards arising from badly maintained electrical equipment, such as soldering equipment, display screen equipment and hand irons, manual handling hazards, fire hazards, chemical hazards and the potential for contracting upper limb disorders. It is common for family members to assist with the work, and in some cases children may be illegally involved in the home working activities as a means of increasing income by the home worker.

The duties of employers

Employers have a general duty of care at common law to ensure that work activities undertaken by employees are safe and without risks to health, and that employees are adequately trained to undertake the work. Under the criminal law, employers have duties under the HASAWA and certain regulations towards home workers comparable with the duties owed to employees working at a specific workplace.

While a home worker's house may not be under the direct control of an employer, the general and specific duties of employers towards employees apply to this type of work. Employers therefore must have some form of procedure for safety monitoring of home workers' work activities with particular reference to their working environment; the tasks undertaken; equipment, articles and substances used; and structural features of the room or area where the work is undertaken. Home workers should also receive appropriate information, instruction, training, and supervision. In the last case, supervision may take the form of regular visits by a home workers' co-ordinator or supervisor to ensure written safety procedures are being observed.

In particular, the employer should prepare and put into practice a statement of policy with respect to home working as part of the organisation's statement of health and safety policy. Such a statement should incorporate the employer's duties towards home workers and the organisation, and arrangements for ensuring these duties are put into practice.

Work area inspections and risk assessment

Before employing people to undertake work at home, employers should ensure that an inspection of the home work area is undertaken, with particular reference to environmental working conditions, the safety of work equipment, including that owned by the home worker, fire safety, the storage of hazardous substances and the extent of manual handling required. Where appropriate, risk assessment should be undertaken with a view to identifying the significant hazards and the preventive and protective measures necessary.

It may be necessary for the employer to designate specific areas where the work must be undertaken.

Many home workers use display screen equipment as a significant part of their work. Under the Health and Safety (Display Screen Equipment) Regulations an employer must undertake a workstation risk analysis in the case of defined display screen equipment 'users', that is, 'employees who habitually use display screen equipment as a significant part of their normal work'. On this basis, workstations may need to be modified to meet the requirements of these regulations.

Similar provisions apply in the case of home workers involved in manual handling operations on a regular basis, where it would be necessary for a manual handling risk assessment to be undertaken by the employer with a view to preventing or controlling risks arising from this work.

LONE WORKING SITUATIONS

What is a lone worker?

A 'lone worker' or 'solitary worker' is defined as anyone who works alone out of contact with other persons. People such as maintenance engineers, post deliverers, housing officers, long-distance drivers and sales representatives might, for example, be classified as lone workers even though they may be in contact with other people, such as customers, subcontractors or suppliers, on a regular basis. In some cases, builders and the employees of building contractors may work on their own at some distance from a main location.

General duties of employers

While there is no specific legal prohibition on lone working, the general and specific duties of employers under the HASAWA apply. Moreover, the employer must, under the MHSWR, asses the risks arising from lone working and plan, organise, control, monitor and review any particular protective measures to ensure that the lone workers are not subjected to more significant risks than other employees who work as a group.

Lone working arrangements

In the design of safe systems of work involving lone working, the following factors should be considered:

(a) careful selection of employees who are fit, competent and reliable;
(b) the need to undertake a suitable and sufficient risk assessment of the lone working activity, which must be kept under review, together with regular monitoring of individual performance;
(c) the employees concerned must be provided with information, instruction and training so that they are quite clear as to all the significant foreseeable risks which may arise and the measures they must take to ensure their own safety and the safety of other persons;
(d) a formally established safe system of work must be operated including, in certain cases, a Permit to Work system, which incorporates a detailed emergency procedure, for example, where an employee is working in a confined space;

(e) suitable and sufficient communication must be maintained, such as the operation of a radio or telephone-based buddy system, central control or electronic monitoring incorporating non-body movement indication/ panic alarm and radio/satellite location, appropriate to the environment in which the lone worker may be working; and
(f) there must be adequate recognition of the more serious consequences for lone workers of fatigue and stress while travelling or undertaking their particular work.

VULNERABLE GROUPS

Certain groups of people, by virtue of their age or physical condition, are considered to be more vulnerable to accidents and ill health at work. These groups include new or expectant mothers, young people and disabled people. The law requires employers to make extra provisions for these groups.

Management of Health and Safety at Work Regulations: Approved Code of Practice

New and expectant mothers

A risk assessment should take account of the risks to new and expectant mothers.

Where the risk assessment identifies risks to new and expectant mothers and these risks cannot be avoided by the preventive and protective measures taken by an employer, the employer will need to:

(a) Alter her working conditions or hours of work if it is reasonable to do so and would avoid the risks. Or if these conditions cannot be met:
(b) Identify and offer her suitable alternative work that is available. Or if that is not feasible:
(c) Suspend her from work. (The Employment Rights Act requires that this suspension should be on full pay. Employment rights are enforced through tribunals.)

Young people

The employer needs to carry out a risk assessment before young people start work and to see where risk remains, taking account of control measures in place. For young workers, the risk assessment needs to pay attention to areas of risk described in Regulation 19(2) (ie beyond physical or psychological capacity, involving exposure to harmful agents, involving exposure to radiation, etc). For several of these areas the employer will need to assess the risks with the control measures in place under other statutory requirements.

When control measures have been taken against these risks and if a significant risk still remains, no child (young worker under compulsory school age) can be employed to do this work. A young worker, above the minimum school leaving age, cannot do this work unless:

(a) it is necessary for his training;
(b) he is supervised by a competent person; and
(c) the risk will be reduced to the lowest level reasonably practicable.

Disabled workers

The Disability Discrimination Act 1995 makes it unlawful for an employer to discriminate against a disabled job applicant or worker with respect to selection for jobs, terms and conditions of employment, promotion or transfer, training, employment benefits and dismissal or other detrimental treatment. A person has a disability for the purposes of the Act if he has a physical or mental impairment which has a substantial and long-term adverse effect on his ability to carry out normal day-to-day activities. A disabled person is a person who has a disability. An employer must make reasonable adjustments where working arrangements and/or the physical features of a workplace cause substantial disadvantage for a disabled person in comparison with those who are not disabled.

Reasonable steps that an employer may need to take, and which may need to be covered in the risk assessment process involving a disabled person, include:

(a) altering working hours;
(b) allowing time off for rehabilitation or treatment;
(c) allocating some of the disabled person's duties to someone else;
(d) transferring the disabled person to another vacancy or another place of work;
(e) giving or arranging training;
(f) providing a reader or interpreter;
(g) acquiring or modifying equipment or reference manuals;
(h) adjusting the premises.

Physical features of the premises which may require adjustment are:

(a) those arising from the design or construction of a building;
(b) exits or access to buildings;
(c) fixtures, fittings, furnishings, equipment or materials;
(d) any other physical element or quality of land or the premises.

Typical examples of physical features that might cause substantial disadvantage to a disabled person are the absence of ramps for wheelchair users, inadequate lighting for someone with restricted vision, doors that are too narrow for wheelchair users and a chair which is unsuitable.

Current trends in health and safety legislation

1. All modern health and safety legislation is largely driven by European Directives. For instance:

   (a) the Directive 'on the health and safety of workers at work' was implemented in the UK as the Management of Health and Safety at Work Regulations 1992; and
   (b) the Temporary and Mobile Construction Sites Directive was implemented in the UK as the Construction (Design and Management) Regulations 1994.

2. Regulations produced since 1992 do not, in most cases, stand on their own. They must be read in conjunction with the general duties imposed on employers under the MHSWR, in particular the duties relating to:

   (a) risk assessment;
   (b) the operation and maintenance of safety management systems;
   (c) the appointment of competent persons;
   (d) establishment and implementation of emergency procedures;
   (e) provision of information which is comprehensible and relevant;
   (f) co-operation, communication and co-ordination between employers in shared workplaces, eg construction sites, office blocks, industrial estates;
   (g) provision of comprehensible health and safety information to employees from an outside undertaking;
   (h) assessment of human capability prior to allocating tasks;
   (i) provision of health and safety training and instruction; and
   (j) specific provisions for pregnant workers and young persons.

DOCUMENTATION AND RECORD KEEPING REQUIREMENTS

Current health and safety legislation places considerable emphasis on the documentation of policies, procedures and systems of work and the maintenance of certain records. The following are some of the documents and records that are required to be produced and maintained:

- Statement of health and safety policy (Health and Safety at Work etc Act 1974).
- Risk assessments in respect of:
  - workplaces (Management of Health and Safety at Work Regulations 1999 and Workplace (Health, Safety at Welfare) Regulations 1992);

- work activities (Management of Health and Safety at Work Regulations 1999 and Workplace (Health, Safety and Welfare) Regulations 1992);
- work groups (Management of Health and Safety at Work Regulations);
- new or expectant mothers (Management of Health and Safety at Work Regulations 1999);
- young persons (Management of Health and Safety at Work Regulations 1999);
- work equipment (Provision and Use of Work Equipment Regulations 1998);
- personal protective equipment (Personal Protective Equipment Regulations 1992);
- manual handling operations (Manual Handling Operations Regulations 1992);
- display screen equipment (Health and Safety (Display Screen Equipment) Regulations 1992);
- substances hazardous to health (Control of Substances Hazardous to Health Regulations 2002);
- significant exposure to lead (Control of Lead at Work Regulations 2002);
- noise levels in excess of 80 dBA (Control of Noise at Work Regulations 2005);
- before a radiation employer commences a new activity involving work with ionising radiation (Ionising Radiations Regulations 1999);
- the presence or otherwise of asbestos in non-domestic premises (Control of Asbestos at Work Regulations 2002);
- where a dangerous substance is or is liable to be present at the workplace (Dangerous Substances and Explosive Atmospheres Regulations 2002).

- Safe systems of work, including permits to work and method statements.
- Pre-tender stage health and safety plan and construction phase health and safety plan (Construction (Design and Management) Regulations 2007).
- Planned preventive maintenance schedules (Workplace (Health, Safety and Welfare) Regulations 1992 and Provision and Use of Work Equipment Regulations 1998).
- Cleaning schedules (Workplace (Health, Safety and Welfare) Regulations 1992).
- Written scheme of examination for specific parts of an installed pressure system or of a mobile system and the last report relating to a system by a competent person (Pressure Systems Safety Regulations 2000).

- Written plan of work identifying those parts of a premises where asbestos is or is liable to be present in a premises and detailing how that work is to be carried out safely and without risk to health (Control of Asbestos at Work Regulations 2002).
- Records of examinations and tests of exhaust ventilation equipment and respiratory protective equipment and of repairs carried out as a result of those examinations and tests (Control of Lead at Work Regulations 1999, Control of Substances Hazardous to Health Regulations 2002 and Control of Asbestos at Work Regulations 2002).
- Record of air monitoring carried out in respect of:
  - specified substances or processes;
  - lead;
  - asbestos;

  (Control of Substances Hazardous to Health Regulations 2002, Control of Lead at Work Regulations 1998 and Control of Asbestos at Work Regulations 2002).
- Record of examination of respiratory protective equipment (Ionising Radiations Regulations 1999).
- Records of air monitoring in cases where exposure to asbestos is such that a health record is required to be kept (Control of Asbestos at Work Regulations 2002).
- Personal health records (Control of Lead at Work Regulations 1999, Ionising Radiations Regulations 1999, Control of Substances Hazardous to Health Regulations 2002 and Control of Asbestos at Work Regulations 2002).
- Personal dose records (Ionising Radiations Regulations 1999).
- Record of quantity and location of radioactive substances (Ionising Radiations Regulations 1999).
- Record of investigation of certain notifiable occurrences involving release or spillage of a radioactive substance (Ionising Radiations Regulations 1999).
- Record of suspected overexposure to ionising radiation during medical exposure (Ionising Radiations Regulations 1999).
- Major accident prevention policy (Control of Major Accident Hazards Regulations 1999).
- Off-site emergency plan (Control of Major Accident Hazards Regulations 1999).
- Declaration of conformity by the installer of a lift and the manufacturer of a safety component for a lift together with any technical documentation or other information in relation to a lift or safety component required to be retained under the conformity assessment procedure (Lifts Regulations 1997).

- Declaration of conformity by the manufacturer of pressure equipment and assemblies (as defined) together with technical documentation or other information in relation to an item of pressure equipment and assemblies required to be retained under the conformity assessment procedure used (Pressure Equipment Regulations 1999).
- Any technical documentation or other information required to be retained under a conformity assessment procedure and a periodic inspection procedure (Transportable Pressure Vessels Regulations 2001).
- Procedures for serious and imminent danger and for danger areas (Management of Health and Safety at Work Regulations 1999).
- Emergency procedure to protect the safety of employees from an accident, incident or emergency related to the presence of a dangerous substance at the workplace (Dangerous Substances and Explosive Atmospheres Regulations 2001).
- Contingency plan in the event of a radiation accident (Ionising Radiations Regulations 1999).
- Local rules in respect of controlled areas and supervised areas (Ionising Radiations Regulations 1999).
- Written arrangements for non-classified persons (Ionising Radiations Regulations 1999).

The employer's duties at common law

Common law (case law) is a special body of law that has developed over the centuries. It is based on the judgements of the courts, each judgement containing a judge's enunciation of the facts, a statement of the law applying to the case and his *ratio decidendi* or legal reasoning for the conclusion or finding that he has arrived at. These judgements are recorded in the various Law Reports and form a body of decisions or precedents which other courts must follow.

At common law, employers owe their employees a general duty to take reasonable care in order to avoid injuries, diseases and deaths occurring at work. In particular, employers must:

(a) provide a safe place of work with safe means of access and egress;
(b) provide and maintain safe appliances and equipment and plant for doing the work;
(c) provide and maintain a safe system of work; and
(d) provide competent people to undertake the work.

(Wilsons & Clyde Coal Co Ltd v English (1938) 2 AER 628)

All these common law duties were incorporated in Section 2 of the HASAWA under the general duties of the employer.

Negligence

The above common law duties feature in the general law of negligence and constitute specific aspects of the duty to take reasonable care. Thus, in order to prove negligence, a claimant must satisfy the following criteria, which are incorporated in the definition of 'negligence', ie:

(a) the existence of a duty of care owed by the defendant to the plaintiff, eg by the employer to the employee;
(b) breach of that duty; and
(c) injury, damage or loss resulting from or caused by that breach.

(Lochgelly Iron & Coal Co Ltd v M'Mullan (1934) AC 1)

Breach of statutory duty

Quite apart from allowing a civil claim under the law relating to negligence, the courts have sometimes recognised that a breach of duty imposed by a statute (or regulations made under a statute) may give rise to a civil claim for damages. This applies where a statute imposes a duty but makes no reference to civil liability for injury, damage or loss caused by breach. In such a case, the approach of the courts has been to ask, '*Was the duty imposed specifically for the protection of a particular class of persons or was it intended to benefit the general public at large?*' If the answer relates to the former, a civil claim may be allowed.

Certain regulations, eg the MHSWR and Construction (Design and Management) Regulations 2007, incorporate an excluding clause, ie excluding claims for civil liability on the basis of breach of statutory duty. Where such an excluding clause is absent, as with the Workplace (Health, Safety and Welfare) Regulations 1992, then, by implication, it may be possible for a claimant to plead breach of statutory duty.

OCCUPIERS' LIABILITY

Occupiers' liability is a branch of civil law concerned with the duties of occupiers of premises to all those who may enter on to those premises. The legislation covering this area of civil liability is the Occupiers Liability Act (OLA) 1957 and, specifically in the case of trespassers, the Occupiers Liability Act 1984.

Under the OLA 1957 an occupier owes a *common duty of care* to all lawful visitors. This common duty of care is defined as:

> 'a duty to take such care as in all the circumstances of the case is reasonable to see that the visitor will be reasonably safe in using the premises for the purposes for which he is invited or permitted by the occupier to be there'.

Section 1 of the Act defines the duty owed by occupiers of premises to all persons lawfully on the premises in respect of 'dangers due to the state of the premises or to things done or omitted to be done on them'.

The Act regulates the nature of the duty imposed in consequence of a person's occupation of premises. The duties are not personal duties but, rather, are based on the occupation of premises, and extend to a person occupying, or having control over, any fixed or movable structure, including any vessel, vehicle or aircraft.

Protection is afforded to all *lawful* visitors, whether they enter for the occupier's benefit, such as customers or clients, or for their own benefit, for instance a police officer, though not to persons exercising a public or private right of way over premises.

Occupiers have a duty to erect notices warning visitors of imminent danger, such as an uncovered pit or obstruction. However, section 2(4) states that a warning notice does not, in itself, absolve the occupier from liability, unless, in all the circumstances, it was sufficient to enable the visitor to be reasonably safe.

Furthermore, while an occupier, under the provisions of the Act, could have excused his liability by displaying a suitable prominent and carefully worded notice, the chance of such avoidance is not permitted as a result of the Unfair Contract Terms Act 1977. This Act states that it is not permissible to exclude liability for death or injury due to negligence by a contract or by a notice, including a notice displayed in accordance with section 2(4) of the Occupiers Liability Act.

Trespassers

A trespasser is defined in common law as a person who:

(a) goes on premises without invitation or permission;
(b) although invited or permitted to be on premises, goes to a part of the premises to which the invitation or permission does not extend;
(c) remains on premises after the invitation or permission to be there has expired;
(d) deposits goods on premises when not authorised to do so.

Section 1 of the OLA 1984 imposes a duty on an occupier in respect of trespassers, namely persons who may have a lawful authority to be in the vicinity or not, who may be at risk of injury on the occupier's premises. This duty can be discharged by issuing some form of warning such as the display of hazard warning notices, but such warnings must be very explicit. It is not good enough, however, merely to display a notice. The requirements of notices must be actively enforced by the occupier.

Generally, the displaying of a notice, the clarity, legibility and explicitness of such a notice, and evidence of regularly reminding people of the message outlined in the notice, may count to a certain extent as part of a defence when sued for injury by a simple trespasser under the Act.

Children

Children, generally, from a legal viewpoint have always been deemed to be less responsible than adults. The OLA 1957 is quite specific on this matter. Section 2(3)(a) requires an occupier to be prepared for children to be less careful than adults. Where, for instance, there is something, or a situation, on the premises that is a lure or attraction to a child, such as a pond, a colony of frogs, an old motor car or a derelict building, this can constitute a trap as far as a child is concerned. Should a child be injured as a result of this trap, the occupier could then be liable. Much will depend upon the location of the premises, for instance, whether or not it is close to houses or a school or is in an isolated location, such as a farmyard deep in the countryside but, in all cases, occupiers must consider the potential for child trespassers and take appropriate precautions.

COURTS AND TRIBUNALS

There are two distinct systems whereby the courts deal with criminal and civil actions respectively. However, some courts have both criminal and civil jurisdiction. The following list outlines the court system operating in the UK.

1. Magistrates court

This is the lowest of the courts in England and Wales and deals mainly with criminal matters. Its jurisdiction is limited. Magistrates determine and sentence for many of the less serious offences. They also hold preliminary examinations into other offences to ascertain whether the prosecution can show a *prima facie* case on which the accused may be committed for trial at a higher court. The Sheriff court performs a parallel function in Scotland, although procedures differ from those of the Magistrates court.

2. Crown Court

Serious criminal charges and cases where the accused has the right to jury trial are heard on indictment in the Crown Court before a judge and jury. This court also hears appeals from magistrates courts.

3. County courts

These courts operate on an area basis and deal in the first instance with a wide range of civil matters. They are limited, however, in the remedies that can be applied. Cases are generally heard by circuit judges or registrars, the latter having limited jurisdiction.

4. High Court of Justice

More important civil matters, because of the sums involved or legal complexity, will start in the High Court of Justice before a High Court judge. The High Court has three divisions:

(a) Queen's Bench – deals with contract and torts;
(b) Chancery – deals with matters relating to areas such as land, wills, bankruptcy, partnerships and companies;
(c) Family – deals with matters involving issues such as adoption of children, marital property and disputes.

In addition, the Queen's Bench Division hears appeals on matters of law:

(a) from the magistrates courts and from the Crown Court on a procedure called 'case stated'; and
(b) from some tribunals, for example the finding of an industrial tribunal on an enforcement notice under the Health and Safety at Work Act.

It also has some supervisory powers over the lower courts and tribunals. These can be exercised if the latter exceed their jurisdiction, fail to undertake their responsibilities properly or neglect to carry out any of their duties.

The High Court, the Crown Court and the Court of Appeal are known as the Supreme Court of Judicature.

5. The Court of Appeal

The Court of Appeal has two divisions:

(a) the Civil Division, which hears appeals from the county courts and the High Court; and
(b) the Criminal Division, which hears appeals from the Crown Court.

6. The House of Lords

The Law Lords deal with important matters of law only. This can only follow an appeal to the House of Lords from the Court of Appeal and in restricted circumstances from the High Court.

7. European Court of Justice

This is the supreme law court, whose decisions on interpretation of European Union law are sacrosanct. Such decisions are enforceable through the network of courts and tribunals in all member states.

Cases can only be brought before this court by organisations, or by individuals representing organisations.

Tribunals

Industrial tribunals were first established under the Industrial Training Act of 1964 to deal with appeals against industrial training levies by employers. They now cover many industrial matters. These include industrial relations issues, cases involving unfair dismissal, equal pay and sex discrimination.

Composition

Each tribunal consists of a legally qualified chairman appointed by the Lord Chancellor and two lay members, one from management and one from a trade union. These representatives are selected from panels maintained by the Department for Employment and Learning following nominations from employers' organisations and trade unions.

Decisions

When all three members of a tribunal are sitting the majority view prevails.

Complaints relating to health and safety matters

Industrial tribunals deal with the following employment/health and safety issues:

(a) appeals against improvement and prohibition notices served by enforcement officers;
(b) time off for the training of safety representatives (Safety Representatives and Safety Committees Regulations 1977, Reg 11 (1)(a));
(c) failure of an employer to pay a safety representative for time off for undertaking his functions and training (Safety Representatives and Safety Committees Regulations 1977, Reg 11(1)(b));
(d) failure of an employer to make a medical suspension payment (Employment Protection (Consolidation) Act 1978, Sec 22); and
(e) dismissal, actual or constructive, following a breach of health and safety law, regulation and/or term of employment contract.

THE HEALTH AND SAFETY INFORMATION FOR EMPLOYEES REGULATIONS 1999

These regulations require information relating to health, safety and welfare to be furnished to employees by means of posters or leaflets in the form approved and published for the purposes of the Regulations by the HSE (Regulations 3 and 4). Copies of the form of poster or leaflets approved in this way may be obtained from HMSO.

The regulations also require the name and address of the enforcing authority and the address of the employment medical advisory service to be written in the appropriate space on the poster (Regulation 5 (1)); and where the leaflet is given the same information should be specified in a written notice accompanying it (Regulation 5 (3)).

SUMMARY

1. The HASAWA places specific duties on employers, employees, occupiers of premises, and designers, manufacturers, importers and suppliers of articles and substances used at work.
2. Inspectors appointed under the Act have powers of entry at any reasonable time and at any time in special circumstances.
3. Inspectors have powers to serve both improvement notices and prohibition notices, to take samples, to require dismantling and testing, and to seize unsafe articles and / or substances.
4. The Safety Representatives and Safety Committees Regulations 1977 provide for the appointment in prescribed cases by recognised trade unions of safety representatives from among the employees. The specific functions of safety representatives are outlined in the regulations.
5. Where requested in writing by at least two safety representatives, the employer must form a safety committee.
6. The employer establishes the role, objectives and functions of the safety committee.
7. Under the Health and Safety (Consultation with Employees) Regulations 1996 employers must consult any employees who are not covered by the Safety Representatives and Safety Committees Regulations 1977.
8. Where an employer employs more than five persons, he must prepare and bring to the notice of employees a statement of health and safety policy.

9. The MHSWR bring in provisions to implement the European Framework Directive. It must be appreciated that all these new duties imposed on employers are of an 'absolute' or strict nature compared with duties under the HASAWA and previous regulations, which are qualified by the term 'so far as is reasonably practicable'. The room for manoeuvre on the part of employers, when charged with an offence under these regulations, has been substantially reduced.
10. The five sets of regulations which accompany the MHSWR, and which should be read in conjunction with these regulations, revoked parts of the old legislation, such as the FA and the OSRPA, and reinforce many of the general requirements of the MHSWR. The details of the requirements of these five sets of regulations are covered in specific chapters.
11. Specific provisions are made for pregnant workers and young persons.
12. Where employees are working away from base, both the employer and host employer have specific responsibilities.

REFERENCES

Health and Safety Commission (1976) *Guidance Notes on Employers' Policy Statements for Health and Safety at Work*, HMSO, London

Health and Safety Commission (1977) *Safety Representatives and Safety Committees Regulations 1977*, HMSO, London

Health and Safety Commission (1977) *Writing Your Health and Safety Policy Statement – How to prepare a safety policy statement for a small business*, HMSO, London

Health and Safety Commission (1999) *Management of Health and Safety at Work Regulations 1999 and Approved Code of Practice*, HMSO, London

Health and Safety Commission (2002) *Directors' Responsibilities for Health and Safety* (Leaflet INDG343), HSE Books, Sudbury

Health and Safety Executive (1980) *Effective Policies for Health and Safety*, HMSO, London

Health and Safety Executive (1991) *Successful Health and Safety Management (HS(G)65)*, HMSO, London

Health and Safety Executive (1999) *Health and Safety Law: What you should know*, HSE Information Centre, Sheffield

Health and Safety Executive (2001) *Health and Safety in Annual Reports: Guidance from the Health and Safety Commission* [Online] HSE website: www.hse.gov.uk/revitalising/annual.htm

Health and Safety Executive (2002) *Health and Safety Regulation: A short guide*, HSE, London

Health and Safety Executive (2003) *Passport Schemes for Health, Safety and the Environment: A good practice guide*, HSE Books, Sudbury, Suffolk [online] http://www.hse.gov.uk/pubns/indg381.pdf

Health and Safety Executive (2009) *Statutory Instruments (Regulations) Owned and Enforced by HSE*, HSE Information Centre

Health and Safety Executive and Institute of Directors (2009) *Leading Health and Safety at Work: Leadership actions for directors and board members*: IND(G)417, HSE Books, Sudbury, Suffolk

Ministry of Justice (2008): *Understanding the Corporate Manslaughter and Corporate Homicide Act 2007*, HMSO, London

Secretary of State for Employment (1972) *Report of the Committee on Safety and Health at Work* (Robens Report) (Cmnd 5034), HMSO, London

Stranks, J (2005) *Health and Safety Law*, Prentice Hall, London

2

Health and Safety Management

WHAT IS MANAGEMENT?

One definition is:

'The effective use of resources in the pursuit of organisational goals.'

'Effective' implies achieving a balance between the risk of being in business and the cost of eliminating or reducing those risks.

Management entails leadership, authority and co-ordination of resources, together with:

(a) planning and organisation;
(b) co-ordination and control;
(c) communication;
(d) selection and placement of subordinates;
(e) training and development of subordinates;
(f) accountability; and
(g) responsibility.

Management resources include:

(a) people – employees, managers;
(b) land and buildings; and
(c) capital; together with
(d) time; and
(e) management skills in co-ordinating the use of resources.

HEALTH AND SAFETY MANAGEMENT

Health and safety management is no different from other forms of management.

It covers:

(a) the management of the health and safety operation at national and local level – planning, organising, controlling, objective setting, establishing accountability and the setting of policy;
(b) measurement of health and safety performance on the part of individuals and specific locations; and
(c) motivating managers to improve standards of health and safety performance in those areas under their control.

Decision making is a very important feature of the management process. This can be summarised thus:

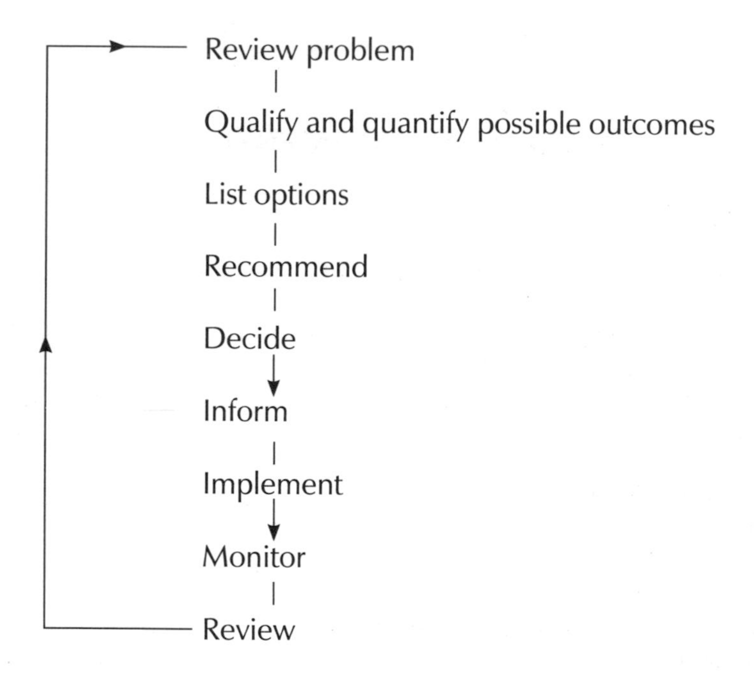

Management performance

Management is concerned with people at all levels of the organisation and human behaviour, in particular human personal factors such as attitude, perception, motivation, personality, learning and training. Communication brings these various behavioural factors together.

The management of health and safety is also concerned with organisational structures, the climate for change within an organisation, individual roles within the organisation and the problem of stress which takes many forms. Several questions must be asked at this stage:

1. Are managers good at managing health and safety?
2. What is their attitude to health and safety? Is it of a proactive or reactive nature?
3. How do their attitudes affect the development of health and safety systems?
4. Is health and safety a management tool, or a problem thrust on to them by the enforcement agencies?

Health and safety management in practice

While health and safety management covers many areas, there are a number of aspects which are significant, namely:

(a) the company statement of health and safety policy;
(b) procedures for health and safety monitoring and performance measurement;
(c) clear identification of the objectives and standards which must be measurable and achievable by the persons concerned;
(d) the system for improving knowledge, attitudes and motivation and for increasing individual awareness of health and safety issues, responsibilities and accountabilities;
(e) procedures for eliminating potential hazards from plant, machinery, substances and working practices through risk assessment, the design and operation of safe systems of work and other forms of hazard control; and
(f) measures taken by management to ensure legal compliance.

SUPERVISORS AND HEALTH AND SAFETY

The supervisor has a significant role in the management of health and safety. The most important features of an effective supervisor in this respect can be summarised thus:

- **Introduction.** Getting to know the employees in his charge, particularly in the case of young persons, and to be known by them.
- **Instruction.** Passing on information and theory in a clear manner with regard to safe systems of work, the correct use of personal protective equipment, accident and hazard reporting procedures, etc.
- **Demonstration.** Actually showing by practical demonstration how a task is done safely.
- **Practice.** Making reasonable allowance for employees to become proficient in tasks, including any precautions necessary to ensure safe working.

- **Monitoring performance.** Observing and measuring employees' extent of proficiency, including compliance with formal safety procedures.
- **Reporting.** Making a fair evaluation of employees' performance for management including their compliance with safety procedures.
- **Correcting and encouraging.** Correcting and encouraging employees as necessary with respect to, for example, safe systems of work.

Supervisor training

All the above factors should be considered in the training of a supervisor, and particularly in terms of his responsibilities and duties for ensuring sound levels of health and safety performance in his section.

SUCCESSFUL HEALTH AND SAFETY MANAGEMENT – HS(G)65

This HSE publication identifies the key elements of successful health and safety management. The Summary to HS(G)65 makes the following points (see also Figure 2.1).

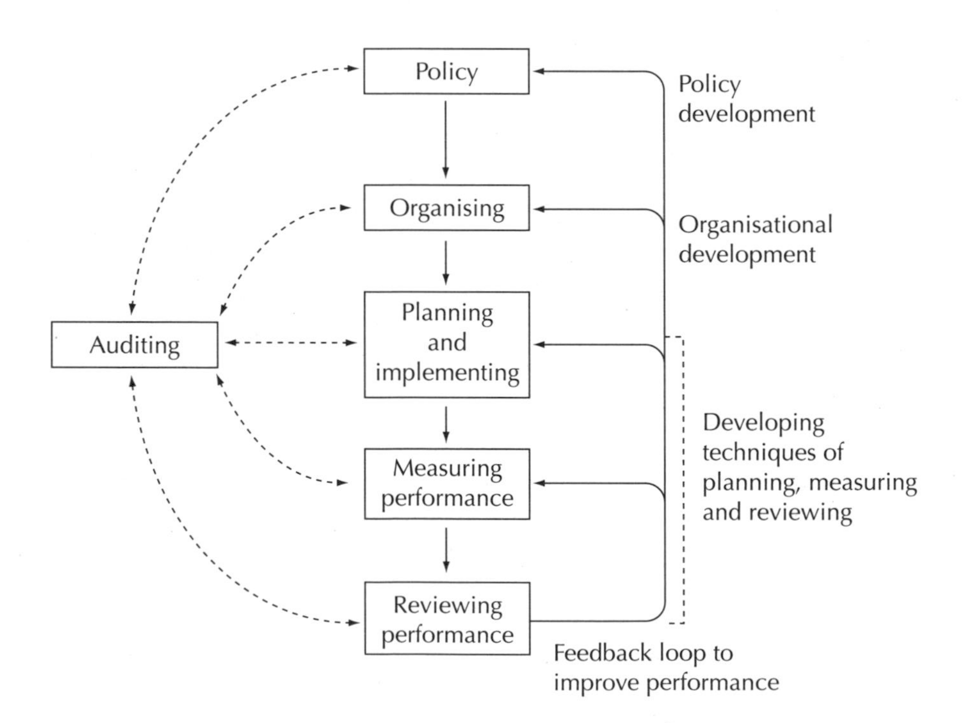

Figure 2.1 Key elements of successful health and safety management (HSE)

Policy

Organisations that are successful in achieving high standards of health and safety have health and safety policies which contribute to their business performance, while meeting their responsibilities to people and the environment in a way that fulfils both the spirit and the letter of the law. In this way they satisfy the expectations of shareholders, employees, customers and society at large. Their policies are cost effective and aimed at achieving the preservation and development of physical and human resources and reductions in financial losses and liabilities. Their health and safety policies influence all their activities and decisions, including those to do with the selection of resources and information, the design and operation of working systems, the design and delivery of products and services, and the control and disposal of waste.

Organising

Organisations that achieve high health and safety standards are structured and operated so as to put their health and safety policies into effective practice. This is helped by the creation of a positive culture that secures involvement and participation at all levels. It is sustained by effective communications and the promotion of competence that enables all employees to make a responsible and informed contribution to the health and safety effort. The visible and active leadership of senior managers is necessary to develop and maintain a culture supportive of health and safety management. Their aim is not simply to avoid accidents, but to motivate and empower people to work safely. The vision, values and beliefs of leaders become the shared 'common knowledge' of all.

Planning

These successful organisations adopt a planned and systematic approach to policy implementation. Their aim is to minimise the risks created by work activities, products and services. They use risk assessment methods to decide priorities and set objectives for hazard elimination and risk reduction. Performance standards are established and performance is measured against them. Specific actions needed to promote a positive health and safety culture and to eliminate and control risks are identified. Wherever possible risks are eliminated by the careful selection and design of facilities, equipment and processes or minimised by the use of physical control measures. Where this is not possible systems of work and personal protective equipment are used to control risks.

Measuring performance

Health and safety performance in organisations that manage health and safety successfully is measured against pre-determined standards. This reveals when and where action is needed to improve performance. The success of action taken to control risks is assessed through active self-monitoring involving a range of techniques. This includes an examination of both hardware (premises, plant and substances) and software (people, procedures and systems), including individual behaviour. Failures of control are assessed through reactive monitoring that requires the thorough investigation of any accidents, ill health or incidents with the potential to cause harm or loss. In both active and reactive monitoring the objectives are not only to determine the immediate causes of sub-standard performance but, more importantly, to identify the underlying causes and the implications for the design and operation of the health and safety management system.

Auditing and reviewing performance

Learning from *all* relevant experience and applying the lessons learnt are important elements in effective health and safety management. This needs to be done systematically through regular reviews of performance based on data both from monitoring activities and independent audits of the whole health and safety management system. These form the basis for self-regulation and for securing compliance with sections 2–6 of the Health and Safety at Work etc Act 1974. Commitment to continuous improvement involves the constant development of policies, approaches to implementation and techniques of risk control. Organisations that achieve high standards of health and safety assess their health and safety performance by internal reference to key performance indicators and by external comparison with the performance of business competitors. They often also record and account for their performance in their annual reports.

QUALITY MANAGEMENT AND HEALTH AND SAFETY – BS 8800: *1996 GUIDE TO OCCUPATIONAL HEALTH AND SAFETY MANAGEMENT SYSTEMS*

BS 8800 offers an organisation the opportunity to review and revise its current occupational health and safety arrangements against a standard that has been developed by industry, commerce, insurers, regulators, trade unions and occupational health and safety practitioners. The standard offers all the essential elements required to implement an effective occupational health and safety management system. It is equally applicable to small organisations as to large, complex organisations.

The aims of the standard are to 'improve the occupational health and safety performance of organisations by providing guidance on how management of occupational health and safety may be integrated with the management of other aspects of the business performance in order to:

(a) minimise risk to employees and others;
(b) improve business performance; and
(c) assist organisations to establish a responsible image in the workplace.'

The benefits

While organisations recognise that they have a duty to manage health and safety, their senior managers may also believe that management of the same is a financial burden which gives very little positive return, similar to VAT and PAYE. This belief frequently results in a lack of, or limited commitment to, health and safety by these managers.

Conversely, those organisations that do subscribe with commitment usually find enormous benefit. Apart from the obvious direct effect on staff morale, they find a positive contribution to the bottom line of their operational costs. The following benefits can be gained:

(a) improved commitment from staff (and positive support from trade unions);
(b) reduction in staff absenteeism;
(c) improved production output, through reductions in downtime from incidents;
(d) reduction in insurance premiums;
(e) improved customer confidence;
(f) reductions in claims against the organisation; and
(g) a reduction in adverse publicity.

Making good progress

The way forward for successful health and safety management is to involve everyone in the organisation, using a proactive approach to identify hazards and to control those risks that are not tolerable. This ensures that those employees at risk are aware of the risks they face and of the need for the control measures.

In order to achieve positive benefits health and safety management should be an integral feature of the undertaking contributing to the success of the organisation. It can be an effective vehicle for efficiency and effectiveness, encouraging employees to suggest improvements in working practices. In an ideal environment health and safety is an agenda item alongside production, services, etc at any senior management review of the undertaking, rather than an inconvenient add-on item.

Status review of the health and safety management system

In any review of an organisation's current health and safety management system, BS 8800 recommends the following four headings:

1. Requirements of relevant legislation dealing with occupational health and safety management issues.
2. Existing guidance on occupational health and safety management within the organisation.
3. Best practice and performance in the organisation's employment sector and other appropriate sectors (eg from relevant HSC's industry advisory committees and trade association guidelines).
4. Efficiency and effectiveness of existing resources devoted to occupational health and safety management.

Statements of health and safety policy

BS 8800 identifies nine key areas that should be addressed in a policy, each of which allows visible objectives and targets to be set, thus:

1. Recognising that occupational health and safety is an integral part of business performance.
2. Achieving a high level of occupational health and safety performance, with compliance to legal requirements as the minimum and continual cost-effective improvement in performance.
3. Provision of adequate and appropriate resources to implement the policy.
4. The publishing and setting of occupational health and safety objectives, even if only by internal notification.
5. Placing the management of occupational health and safety as a prime responsibility of line management, from most senior executive to the first-line supervisory level.
6. Ensuring understanding, implementation and maintenance of the policy statement at all levels in the organisation.
7. Employee involvement and consultation to gain commitment to the policy and its implementation.
8. Periodic review of the policy, the management system and audit of compliance to policy.
9. Ensuring that employees at all levels receive appropriate training and are competent to carry out their duties and responsibilities.

The models

There are two recommended approaches depending on the organisational needs of the business and with the objective that the approach will be integrated into the total management system. One is based on *Successful Health and Safety Management* HS(G)65 (see earlier in this chapter) and the other on ISO 14001 (see Figure 2.2). ISO 14001 is compatible with the environmental standard.

The structure of this approach is such that interfacing or integration with environmental management is relatively straightforward.

For any new system to succeed there must be commitment at the highest management levels, with the plan for implementing such a system underwritten by the board. Planning improvements in health and safety performance should take a proactive approach identifying the risks and the immediate, short, medium and long-term actions required, as opposed to the commonly encountered reactive approach based on the analysis of accident causes and the preparation of statistical information.

Any proactive approach should:

(a) specify the actions necessary and the criteria for assessing satisfactory implementation of same;
(b) set deadlines for completion of the actions necessary; and
(c) identify the persons responsible for ensuring these actions are taken.

The standard specifies a staged approach for developing and implementing a plan, incorporating key stages. (See Figure 2.3.)

The key objectives for Stage 1 are identified from the initial or status review and by undertaking risk assessments and identifying the legal requirements to which the organisation subscribes. This stage identifies the long-term objectives of the organisation. Key objectives dealing with the most significant risks should be identified. These objectives should be both measurable and achievable and, where possible, their implementation should involve employees directly in order to promote the programme and gain their commitment.

Involvement and consultation

In order to gain commitment to the process, employees can be involved in a number of ways, such as training certain members to undertake risk assessments and undertaking safety tours of the workplace to identify aspects of non-conformance, eg failure to wear personal protective equipment or use specified protective measures.

Planning requirements

The principal requirements for planning under BS 8800 are:

(a) *Risk assessment* The organisation should undertake risk assessments, including the identification of hazards.

(b) *Legal and other requirements* The organisation should identify the legal requirements, in addition to the risk assessment, applicable to it and also any other requirements to which it subscribes that are applicable to occupational health and safety management.

(c) *Occupational health and safety management* The organisation should make arrangements to cover the following key areas:
   - (i) overall plans and objectives, including personnel and resources, for the organisation to achieve its policy;
   - (ii) have or have access to sufficient occupational health and safety knowledge, skills and experience to manage its activities safely and in accordance with legal requirements;
   - (iii) operational plans to implement arrangements to control risks identified and to meet the requirements identified;
   - (iv) planning for operational control activities;
   - (v) planning for performance measurement, corrective action, audits and management reviews; and
   - (vi) implementing corrective actions shown to be necessary.

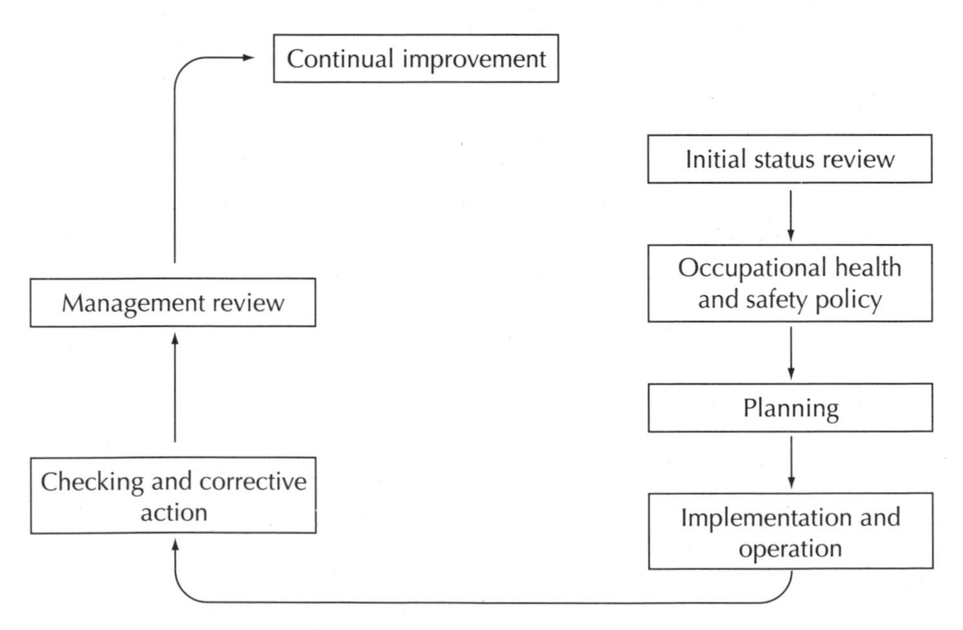

Figure 2.2 An approach based on ISO 14001 – the Plan–Do–Check–Act cycle

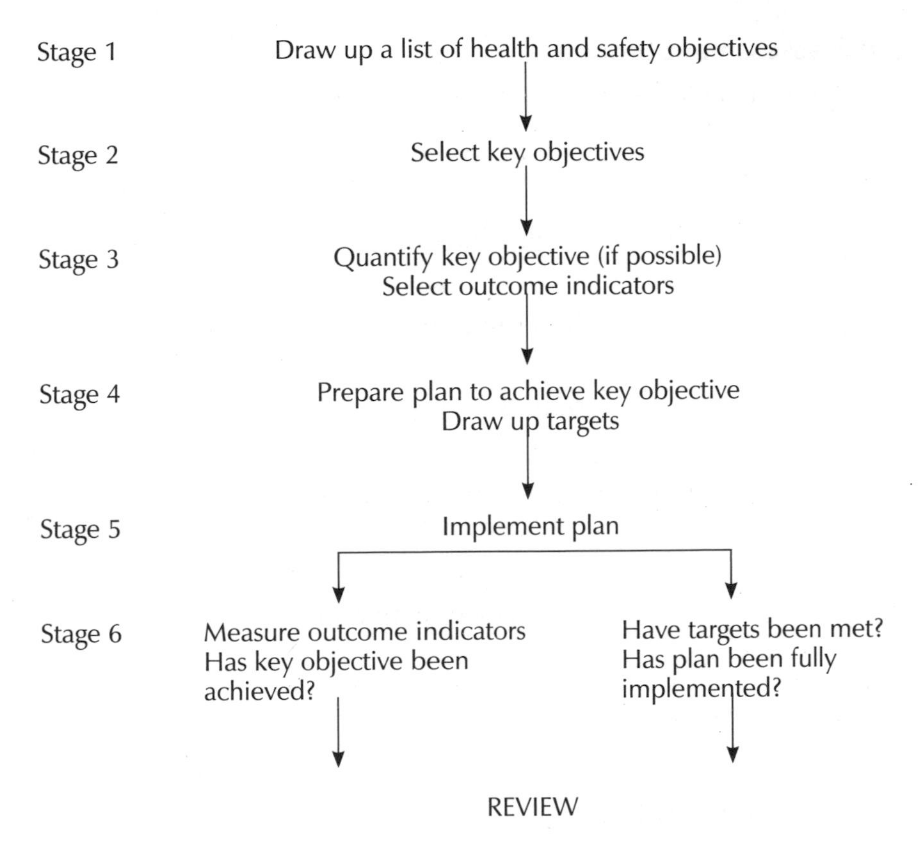

Figure 2.3 The planning process
Note: This diagram covers both the planning and implementation stages to indicate the complete process. Planning involves Stages 1–4.

Risk assessment

The MHSWR impose an absolute duty on an employer to undertake a 'suitable and sufficient assessment' of the risks to his employees and other persons who may be affected by the conduct by him of his undertaking for the purpose of identifying the measures he needs to take to comply with the requirements and prohibitions imposed by or under the relevant statutory provisions.

Having carried out the risk assessment an employer must introduce effective systems to ensure the effective planning, organisation, control, monitoring and review of the preventive and protective measures arising from the risk assessment.

Employees should be involved in the risk assessment process as, in most cases, they will be aware of the hazards arising from work activities.

The risk assessment process

Risk assessment fundamentally takes place in a series of stages:

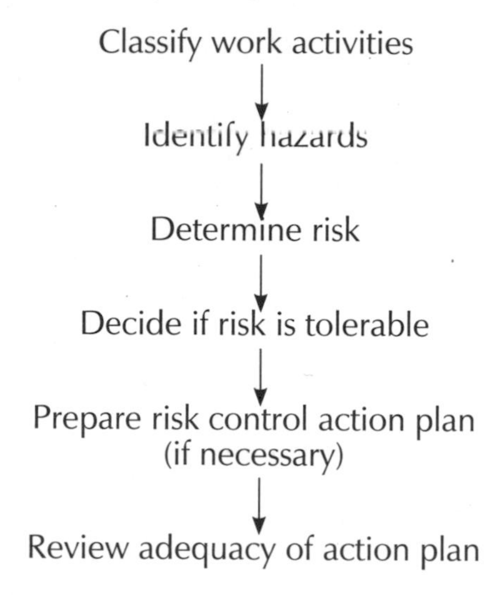

Note: Tolerable risk means that the risk has been reduced to the lowest level that is reasonably practicable.

Effective implementation

There must be total commitment from the board and senior management downwards for effective implementation and operation. Many organisations successfully ensure this by the appointment of a director or executive with responsibility for occupational health and safety and with particular responsibility for the regular review of performance across the organisation. This may be accompanied by a mission statement or other document outlining the commitment of the organisation to the continual improvement of health and safety.

In addition to ensuring that everyone recognises his individual and collective responsibilities, BS 8800 specifies a number of key issues that must be addressed:

(a) structure and responsibility;
(b) training, awareness and competence;
(c) communication;
(d) occupational health and safety management system documentation;
(e) document control;
(f) operational control;
(g) emergency preparedness and response.

Monitoring

This involves the key areas of checking, correcting and auditing.

1. Checking and correcting

Performance measurement is an important feature of the management process and a key way to provide information on the effectiveness of the health and safety management system. *Qualitative* and *quantitative* measures should be considered with the aim of ensuring that proactive systems are recognised as the prime means of control. Examples of proactive monitoring data include:

(a) the extent to which objectives and targets have been set;
(b) the extent to which objectives and targets have been met;
(c) employee perception of management's commitment;
(d) the adequacy of communication of the statement of health and safety policy and other 'core' documents;
(e) the extent of compliance with the 'relevant statutory provisions' and voluntary codes, standards and guidance;
(f) the number of risk assessments carried versus those actually required;
(g) the extent of compliance with risk controls;
(h) the time taken to implement actions on complaints or recommendations;
(i) the frequency of inspections, audits and other forms of monitoring.

BS 8800 identifies three other key areas to be addressed apart from monitoring and measurement: corrective action, records and audit.

2. Auditing

There must be an effective audit system that entails a critical appraisal of all the elements of the management system. This is best undertaken through a form of ongoing audit programme. The process given in BS 8800 for establishing an audit programme is shown below:

Health and safety audits should 'be conducted by persons who are competent and as independent as possible from the activity that is being audited, but may be drawn from within the organisation'.

The auditing process should include:

(a) structured interviews being carried out with key personnel and others to determine that sound procedures are in place, understood and followed;
(b) relevant documentation being examined, such as statements of health and safety policy, risk assessments, written instructions and procedures, health and safety manuals, etc;
(c) inspections being undertaken to confirm the documented procedures and any statements made;
(d) the analysis and interpretation of data such as accidents, near misses, etc.

Management review

A periodic management review is an essential component of the management system. It differs significantly from the audit but draws on the audit, among other indicators, to determine the robustness of the health and safety management system.

BS 8800 suggests that reviews should consider:

(a) the overall performance of the occupational health and safety management system;
(b) the performance of the individual elements of the system;
(c) the findings of audits;
(d) internal and external factors, such as changes in organisational structure, legislation pending, the introduction of new technology, etc,

and identify what action is necessary to remedy any deficiencies.

BENCHMARKING

A benchmark is a reference point which is commonly used in surveying practice. More recently, the term has been used to imply some form of standard against which an organisation can measure performance, and as such it is an important business improvement tool in areas such as quality management. Health and safety benchmarking follows the same principles, whereby an organisation's health and safety performance is compared with a similar organisation or 'benchmarking partner'.

The HSE publication *Health and Safety Benchmarking: Improving together* (IND G301/1999) defines health and safety benchmarking as:

a planned process by which an organisation compares its health and safety processes and performance with others to learn how to:

- reduce accidents and ill health;
- improve compliance with health and safety law; and / or
- cut compliance costs.

The benchmarking process

Health and safety benchmarking is a five-step cycle aimed at ensuring continuous improvement. At the commencement of the process, it would be appropriate to form a small benchmarking team or group, perhaps comprising a senior manager, health and safety specialist, line managers, employee representatives and representatives from the benchmarking partner or trade association.

Step 1 – Deciding what to benchmark

Benchmarking can be applied to any aspect of health and safety, but it is good practice to prioritise in terms of high hazard and risk areas, such as with the use of hazardous substances, with certain types of workplace or working practice. Feedback from the risk assessment process and accident data should identify these priorities. Consultation with the workforce should take place at this stage, together with trade associations that may have experience of the process.

Step 2 – Deciding where you are

This stage of the exercise is concerned with identifying the current level of performance in the selected area for consideration and the desired improvement in performance. Reference should be made at this stage to legal requirements, such as regulations, to ACOPs and HSE guidance on the subject, and to any in-house statistical information. It may also be appropriate to use an audit and / or questionnaire approach to measure the current level of performance.

Step 3 – Selecting partners

In large organisations it may be appropriate to select partners both from within the organisation, perhaps at a different geographical location (internal benchmarking) and from outside the organisation (external benchmarking). With smaller organisations, trade associations or the local Chamber of Commerce may be able to assist in the selection of partners. Local benchmarking clubs operate in a number of areas. Reference should be made to the *Benchmarking Code of Conduct* to ensure compliance with it at this stage.

Step 4 – Working with your partner

With the right planning and preparation, this stage should be straightforward. Any information that is exchanged should be comparable, confidentiality should be respected, and all partners should have a good understanding of the other partners' process, activities and business objectives.

Step 5 – Acting on the lessons learnt

Fundamentally, the outcome of any benchmarking exercise is to learn from other organisations, to learn more about the individual organisation's performance compared with working partners, and to take action to improve performance.

According to the HSE, any action plan should be 'SMARTT', that is:

- **s**pecific;
- **m**easurable;
- **a**greed;
- **r**ealistic;
- **t**rackable;
- **t**imebound.

As with any action plan, the health and safety plan should identify a series of recommendations, the member(s) of the organisation responsible for implementing these recommendations and a timescale for their implementation. Progress in implementation should be monitored on a regular basis. In some cases, it may be necessary to redefine objectives in the light of, for example, recent new legislation. There should be continuing liaison with benchmarking partners during the various stages of the action plan.

Pointers to success

To succeed in health and safety benchmarking, there should be:

- senior management resources and commitment;
- employee involvement;
- a commitment to an open and participative approach to health and safety, including a willingness to share information with others within and outside the organisation;
- comparison with data on a meaningful 'apples with apples' basis;
- adequate research, planning and preparation.

EMERGENCY PROCEDURES

Regulation 7 of the MHSWR requires employers to establish 'procedures for serious and imminent danger and for danger areas'. A 'danger area' is defined in the ACOP as a work environment that must be entered by an employee where the level of risk is unacceptable without special precautions being taken.

Identifying the risks

The risk assessment required under Regulation 3 of the MHSWR should identify the significant risks arising out of work. These could include, for instance, the potential for a major escalating fire, explosion, building collapse, pollution incident, bomb threat and some of the scheduled dangerous occurrences listed in Reporting of Injuries, Diseases and Dangerous Occurrences Regulations 1995 (RIDDOR), eg the explosion, collapse or bursting of any closed pressure vessel. All these events could result in a major incident, which can be defined as one that may:

(a) affect several departments within an undertaking;
(b) endanger the surrounding communities;
(c) be classed as a dangerous occurrence under RIDDOR; or
(d) result in adverse publicity for the organisation with ensuing loss of public confidence and marketplace image.

Fundamentally, the question must be asked: '*What are the worst possible types of incident that could arise from the process or undertaking?*'

Once these major risks, which could result in serious and imminent danger, have been identified, a formal emergency procedure must be produced.

Approved Code of Practice

The ACOP, which should be read in conjunction with the regulations, raises a number of important points with regard to the establishment of emergency procedures:

1. The aim must be to set out clear guidance on when employees and others at work should stop work and how they should move to a place of safety.
2. The risk assessment should identify the foreseeable events that need to be covered by these procedures.
3. Many workplaces or work activities will pose additional risks. All employers should consider carefully in their risk assessment whether such additional risks might arise.
4. The procedures may need to take account of responsibilities of specific employees, eg in the shutting down of plant.
5. The procedures should set out the role, identity and responsibilities of the competent persons nominated to implement the detailed actions.

6. Where specific emergency situations are covered by particular regulations, procedures should reflect any requirements laid on them by these regulations.
7. The procedure should cater for the fact that emergency events can occur and develop rapidly, thus requiring employees to act without waiting for further guidance.
8. Emergency procedures should normally be written down, clearly setting out the limits of action to be taken by all employees.
9. Work should not be resumed after an emergency if a serious danger remains. Consult the emergency authorities if in doubt.
10. In shared workplaces separate emergency procedures should take account of others in the workplace and, as far as is appropriate, should be co-ordinated.

Devising an emergency procedure

The following matters must be taken into account when establishing an emergency procedure:

(a) liaison with external authorities and other local companies;
(b) the appointment of an emergency controller;
(c) the establishment of an emergency control centre;
(d) individuals responsible for initiating the procedure;
(e) notification to local authorities;
(f) call-out arrangements for key personnel;
(g) immediate action on site;
(h) evacuation procedures;
(i) access to records;
(j) public relations arrangements – dealing with the media;
(k) catering and temporary shelter arrangements;
(l) contingency arrangements; and
(m) training, including the participation of external services.

RISK MANAGEMENT

It is common practice for organisations to seek from an insurance company total or blanket cover for all foreseeable risks. However, by the adoption and implementation of good standards of risk management, many of these foreseeable risks could be eliminated or controlled, with the result that only the residual risks, those that cannot be eliminated or controlled through in-house procedures, are actually covered by insurance.

What risk management is

Risk management is defined as 'the elimination or minimisation of the adverse effects of pure and speculative risks to which an organisation is exposed'. Within a business, there are *pure risks*, which can only result in a loss to the organisation. Typical examples include those arising from fire hazards caused by poor housekeeping, or the risk of boiler explosions as a result of inadequate maintenance of boilers and their safety devices. *Speculative risks*, on the other hand, can result in either gain or loss. This form of risk is associated with many financial transactions, such as the acquisition of shares in companies.

Within the context of a risk management programme, risk may be defined as 'the chance of loss', and any programme must therefore be geared to the safeguarding of the organisation's assets, such as employees, materials, machinery, methods, manufactured goods and money.

The role of risk management

The role of risk management in commerce and industry is to:

- consider the impact of certain risky events on the performance of the organisation;
- devise alternative strategies for controlling these risks and / or their impact on the organisation;
- relate these alternative strategies to the general decision framework used by the organisation.

The risk management process

This process incorporates the following elements.

Identification of the exposure to risk

This part of the process entails a survey of all areas within the organisation's activities that could be a source of fortuitous risk. It requires knowledge of:

- all the assets, both tangible and intangible;
- the sources of direct and indirect earnings;
- the liabilities imposed under both common and statute law.

Assets, earnings and legal requirements are all subject to constant change. On this basis the identification of exposure to risk is a constant process of prediction, evaluation and re-evaluation in the light of current conditions and circumstances.

Analysis and evaluation of the risks

A number of questions must be asked at this stage:

- How frequently will the loss occur?
- What is the predicted severity of loss?
- What are the legal and practical implications of a major incident, such as the death of an employee, a major escalating fire or a pollution incident?
- What is the likelihood of the loss arising?
- What particular group of employees or members of the public could be affected by the incident?

The answers to these questions would be based on, for instance, past experience of similar incidents and potential loss situations, simulation exercises and quantified risk assessment techniques. All analysis and evaluation, however, features a strong element of prediction.

Prevention or control of the risk

Deciding on the appropriate prevention and/or control strategy is the heart of the risk management process. Far too often, however, this is left to chance or decided by default, particularly where capital expenditure may be required to prevent or control the exposure to risk.

Many people are involved in risk prevention or control, for example the engineering manager to avoid plant failure, the production manager to ensure products meet specifications, the company legal executive to protect the organisation from criminal prosecution and civil liability, the accountant to protect financial assets from fraud, the security manager to protect the organisation's trade secrets, and the health and safety adviser to ensure compliance with the relevant health and safety procedures and safety requirements. Far too often, however, the efforts of the above persons are unco-ordinated, isolated and fragmented, resulting in confusion among managers about where the true responsibilities and accountabilities for preventing or controlling risk exposures lie. Furthermore, action may only be taken in response to loss-producing incidents, rather than prior to these incidents arising.

The need for teamwork

Teamwork is, therefore, one of the important features of a co-ordinated risk management programme. The prevention and control of risk must be approached in anticipation of future hazards, and the risks arising from them, instead of reacting to incidents after they have occurred. The benefits can be tremendous, in terms of reduced insurance premiums, reduced accident and ill-health costs, reduced fines and civil claims, together with a wide range of indirect benefits, such as improved industrial relations, quality and consultation.

Techniques of risk management

Risk management involves the identification, evaluation and economic control of risks within an organisation.

Risk identification

This is achieved by a multiplicity of techniques, including physical workplace inspections, management and worker discussions, safety audits, job safety analysis and HAZOP studies. It can also involve the study of past accidents to identify areas of high risk.

Typical risks include those associated with:

- fire, flood, storm, impact, explosion, subsidence and other hazards;
- accidents and the use of faulty products;
- error, resulting in loss through damage or malfunction arising from incorrect or mistaken operation of equipment or wrong operation of an industrial programme;
- theft and fraud;
- contravention of social or environmental legislation;
- political risks, such as the appropriation of foreign assets by local governments, or the creation of barriers to the repatriation of overseas profits;
- computer fraud, viruses and espionage;
- product tampering situations;
- malicious damage.

Risk evaluation (or measurement)

This may be based on economic, social or legal considerations:

- **Economic considerations** should include the financial impact on the organisation of the uninsured costs of accidents, the effect on insurance premiums, the overall effect on the profitability of the organisation and the possible loss in production following, for instance, enforcement action by the authorities.
- **Social and humanitarian considerations** should include the general well-being of employees, and the interaction with the general public who either live near the organisation's premises or come into contact with the organisation's operations, such as those affected by transport, noise or effluents, and consumers of the organisation's products or services.
- **Legal considerations** should include possible constraints arising from compliance with health and safety legislation and other legislation concerning fire prevention, pollution and product liability.

The probability and frequency of each occurrence, and the severity of the outcome, including an estimation of the maximum potential loss, will also need to be incorporated in any meaningful evaluation of risk.

Types of risk

There are fundamentally seven areas of risk, all of which are considered to be of increasing concern:

(a) environment-related risk;
(b) security of property and information;
(c) safety of employees and members of the public;
(d) credit risks including the potential for credit fraud;
(e) directors' and officers' liability;
(f) due diligence (for acquisition);
(g) product liability (particularly in the case of tampering).

Risk management strategies

Risk avoidance

This strategy involves a specific management decision to avoid completely a particular risk by terminating the procedure, process or operation that produces the risk, such as the use of a particular hazardous substance in the light of recent knowledge about that substance. Such a decision would be based on identification of the risk and its subsequent evaluation. A further risk avoidance strategy would be, for example, the decision to cease the use of any form of asbestos, replacing it with a safer alternative.

Risk retention

In this case, the risk is retained within the organisation and any consequent loss is financed from within the organisation. Two areas of risk retention must be considered in this case, namely risk retention with knowledge and risk retention without knowledge.

Risk retention with knowledge

In this case a decision is made to meet any resulting loss from within the organisation's resources. Much will depend on the nature of the undertaking's operations in deciding on the risks which will be retained.

Risk retention without knowledge

This may arise as a result of ignorance of the existence of a particular risk, or conversely a failure to insure against it. This situation commonly arises where the risks have not been identified in the first place, or if they have been identified, they have not been correctly measured.

Risk transfer

Risk transfer is the process of legally assigning the costs of certain potential losses from one party to another. The most common way of implementing transfer is some form of insurance. Through the issue of an insurance policy, the insurer undertakes to compensate the insured against losses arising with respect to those risks referred to in the insurance policy.

Risk reduction

Risk management systems incorporate risk reduction through the implementation of a formally established loss control programme directed at protecting the company's assets arising from accidental loss. This programme is subject to regular review and modification where, for instance, new risks are identified and evaluated.

Risk reduction strategies take two stages. The first involves collection of data on as many loss-producing incidents as possible, thereby providing information on which an effective programme of remedial action can be based. Well-controlled systems for the investigation of all loss-producing incidents are crucial to the success of this part of the programme. The second stage is the bringing together and analysis of all areas where losses have arisen from these incidents, and the formulation of future strategies based on information derived from data analysis with the aim of reducing losses and wastage of the organisation's assets.

Risk management programmes

The principal aim of a risk management programme, which is generally run in conjunction with an organisation's insurers, is to provide a cost-effective system designed to protect the resources of an organisation by controlling the risks that it faces. Risk management programmes operate on a phased basis and incorporate the four risk control strategies outlined above. The programme should be monitored and reviewed on a regular basis to ensure objectives are being met, new risks have been identified and incorporated in the programme, and those running the programme are competent to do so.

INFORMATION SOURCES FOR HEALTH AND SAFETY

Formal (primary) sources

1. EU Directives

These are the Community instrument of legislation. Directives are legally binding on the governments of all member states who must introduce national legislation, or use administration procedures where applicable, to implement its requirements.

2. Acts of Parliament (Statutes)

Acts of Parliament can be innovatory, ie introducing new legislation, or consolidating, ie reinforcing, with modifications, existing law. Statutes empower the Minister or Secretary of State to make regulations (delegated or subordinate legislation). Typical examples are the HASAWA and the Factories Act 1961.

3. Regulations (Statutory Instruments)

Regulations are more detailed than the parent Act, which lays down the framework and objectives of the system. Specific details are incorporated in regulations made under the Act, eg the Control of Substances Hazardous to Health (COSHH) Regulations 2002, which were passed pursuant to the HASAWA. Regulations are made by the appropriate Minister or Secretary of State whose powers to do so are identified in the parent Act.

4. Approved Codes of Practice

The HSC is empowered to approve and issue Codes of Practice for the purpose of providing guidance on health and safety duties and other matters laid down in statute or regulations. A Code of Practice can be drawn up by the Commission or the HSE. In every case, however, the relevant government department, or other body, must be consulted beforehand and approval of the Secretary of State must be obtained. Any Code of Practice approved in this way is an Approved Code of Practice (ACOP).

5. Case law (common law)

Case law is an important source of information. It is derived from common law because, traditionally, judges have formulated rules and principles of law as the cases occur for decision before the courts.

What is important in a case is the *ratio decidendi* (the reason for the decision). This is binding on courts of equal rank who may be deciding the same point of law. *Ratio decidendi* is the application of such an established principle to the facts of a given case, for instance, negligence consists of omitting to do what a 'reasonable man' would do in order to avoid causing injury to others.

Case law is found in law reports, for example, the All England Law Reports, the Industrial Cases Reports, the current *Law Year Book* and in professional journals, eg *Law Society Gazette, Solicitors Journal*. In addition, many newspapers carry daily law reports, eg *The Times*, the *Financial Times*, the *Daily Telegraph* and the *Independent*.

The supreme law court is the European Court of Justice, whose decisions are carried in *The Times*.

Secondary sources

A wide range of non-legal sources of information are available, some of which may be quoted in legal situations, however.

1. HSE series of Guidance Notes

Guidance Notes issued by the HSE have no legal status. They are issued on a purely advisory basis to provide guidance on good health and safety practices, specific hazards, etc. There are five series of Guidance Notes – General, Chemical Safety, Plant and Machinery, Medical, and Environmental Hygiene.

2. British Standards

These are produced by the British Standards Institute. They provide sound guidance on numerous issues and are frequently referred to by enforcement officers as the correct way of complying with a legal duty.

British Standard EN1SO 12001 'Safety of machinery' is commonly quoted in conjunction with the duties of employers under the Provision and Use of Work Equipment Regulations 1998 to provide and maintain safe work equipment.

3. Manufacturers' information and instructions

Under Section 6 of the HASAWA (as amended by the Consumer Protection Act 1987) manufacturers, designers, importers and installers of 'articles and substances used at work' have a duty to provide information relating to the safe use, storage, etc of their products. Such information may include operating instructions for machinery and plant and hazard data sheets in respect of dangerous substances. Information provided should be sufficiently comprehensive and understandable to enable a judgement to be made on their safe use at work.

4. Safety organisations

Safety organisations, such as the Royal Society for the Prevention of Accidents (RoSPA) and the British Safety Council, provide information in the form of magazines, booklets and videos on a wide range of health and safety-related topics.

5. Professional institutions and trade associations

Many professional institutions, such as the Institute of Occupational Safety and Health (IOSH), the Chartered Institute of Environmental Health (CIEH) and the British Occupational Hygiene Society (BOHS), provide information, both verbally and in written form. Similarly, a wide range of trade associations provide this information to members.

6. Insurance companies

Most insurance companies provide an information service to clients on a wide range of health and safety-related matters.

7. Department for Work and Pensions

This government department publishes annual statistics on claims for industrial injuries benefit and other matters.

8. Published information

This takes the form of textbooks, magazines, law reports, updating services, microfiche systems, films and videos on general and specific topics.

Internal sources

There are many sources of information available within organisations. These include:

1. Existing written information

This may take the form of statements of health and safety policy, company health and safety codes of practice, specific company policies, eg on the use and storage of dangerous substances, current agreements with trade unions, company rules and regulations, methods, operating instructions, etc.

Such documentation could be quoted in a court of law as an indication of the organisation's intention to regulate activities in order to ensure legal compliance. Evidence of the use of such information in staff training is essential here.

2. Work study techniques

Included here are the results of activity sampling, surveys, method study, work measurement and process flows.

3. Job descriptions

A job description should incorporate health and safety responsibilities and accountabilities. It should take account of the physical and mental requirements and limitations of certain jobs and any specific risks associated with the job. Representations from operators, supervisors and trade union safety representatives should be taken into account.

Compliance with health and safety requirements is an implied condition of every employment contract, breach of which may result in dismissal or disciplinary action by the employer.

4. Accident and ill-health statistics

Statistical information on past accidents and sickness may identify unsatisfactory trends in operating procedures which can be eliminated at the design stage of safe systems of work.

The use of accident statistics and rates, eg accident incidence rate, as a sole measure of safety performance is not recommended, however, due to the variable levels of accident reporting in work situations. Under-reporting of accidents, common in many organisations, can result in inaccurate comparisons being made between one location and another.

5. Task analysis

Information produced by the analysis of tasks, such as the mental and physical requirements of a task, manual operations involved, skills required, influences on behaviour, hazards specific to the task, and learning methods necessary to impart task knowledge must be taken into account.

Job safety analysis, a development of task analysis, will provide the above information prior to the development of safe systems of work.

6. Direct observation

This is the actual observation of work being carried out. It identifies interrelationships between operators, hazards, dangerous practices and situations and potential risk situations. It is an important source of information in ascertaining whether, for instance, formally designed safe systems of work are being operated or safety practices, imparted as part of former training activities, are being followed.

7. Personal experience

People have their own unique experience of specific tasks and the hazards which those tasks present.

The experiences of accident victims, frequently recorded in accident reports, are an important source of information. Feedback from accidents is crucial in order to prevent repetition of these accidents.

8. Incident recall

This is a technique used in a damage control programme to gain information about near-miss accidents.

9. Product complaints

A record of all product complaints and action taken should be maintained. Such information provides useful feedback in the modification of products and the design of new products. Product complaints may also result in action by the enforcement authorities under Section 6 of the HASAWA and / or civil proceedings in the event of injury, damage or loss sustained as a result of a defective product or defect in a product.

SUMMARY

1. There is an absolute duty on employers to manage health and safety at work.
2. 'Management' is defined as the effective use of resources in the pursuit of organisational goals.
3. Communication is an essential feature of the management process.
4. There is more to successful health and safety management than merely complying with current legislation.
5. Health and safety management covers a wide range of issues, including documentation of procedures and systems, monitoring performance, and the implementation of systems for improving knowledge, attitudes and motivation.
6. In order to be able to manage health and safety effectively, managers must be aware of current information sources.
7. Increasingly, employers will be required to show evidence of operating formally established health and safety management systems.

REFERENCES

British Standards Institute (1996) *Guide to Occupational Health and Safety Management Systems: BS8800*, BSI, Milton Keynes

Health and Safety Commission (1999) *Management of Health and Safety at Work Regulations 1999 and Approved Code of Practice*, HMSO, London

Health and Safety Executive (1991) *Successful Health and Safety Management (HS(G)65)*, HMSO, London

Health and Safety Executive (1999) *An Introduction to Health and Safety (IND9259)*, HSE Books, Sudbury

Health and Safety Executive (1999) *Health and Safety Benchmarking: Improving together*, (IND G301), HSE Books, Sudbury

Stranks, J (1994) *Management Systems for Safety*, Pitman, London

3

Reporting, Recording and Investigation of Injuries, Diseases and Dangerous Occurrences

The reporting and investigation of work-related injuries, occupational diseases and scheduled 'dangerous occurrences,' such as a boiler explosion or scaffold collapse, has been a legal requirement for many years. Much of the current legislation is based on the lessons learnt as a result of investigation of these events by the enforcement agencies, government departments and following public enquiries.

All accidents have both indirect and direct causes. They may be associated with unsafe working practices, inadequate supervision and training, human error, poor machinery specification and a host of other causes. (See Figure 3.1 on page 91.)

Moreover, all accidents, no matter how trivial they may appear, cases of occupational disease, and the type of dangerous occurrence mentioned earlier, represent substantial losses to an organisation. It is essential, therefore, that organisations learn by their mistakes with a view to preventing recurrences of these costly losses.

REPORTING REQUIREMENTS

The Reporting of Injuries, Diseases and Dangerous Occurrences Regulations 1995 (RIDDOR) cover the requirement to *notify and report* certain categories of injury and disease sustained by people at work, together with specified dangerous occurrences and gas incidents to the relevant enforcing authority, ie HSE or local authority. The majority of duties on 'responsibile persons' (as defined) are of an absolute nature.

Principal requirements of RIDDOR

1. Responsible person to *notify* the relevant enforcing authority by the quickest practicable means and subsequently make a *report* within 10 days on the approved form in respect of *death*, any defined *major injury* and *dangerous occurrence* arising out of or in connection with work (Regulation 3).
2. Duty on an employer to report the *death* of an employee where, as a result of an accident at work, the injured employee dies within one year of the accident (Regulation 4).
3. Duty on an employer to report cases of *disease* to employees listed in the schedule (Regulation 5).
4. Duties on certain persons to report gas incidents (Regulation 6).
5. Responsible person to keep records of reportable injuries, diseases and dangerous occurrences (Regulation 7).

Notifiable and reportable major injuries

These are listed in Schedule 1 of RIDDOR, thus:

1. Any fracture, other than to the fingers, thumbs or toes.
2. Any amputation.
3. Dislocation of the shoulder, hip or knee.
4. Loss of sight (whether temporary or permanent).
5. A chemical or hot metal burn to the eye or any penetrating injury to the eye.
6. Any injury resulting from electric shock or electrical burn (including any electrical burn caused by arcing or arcing products) leading to unconsciousness or requiring resuscitation or admittance to hospital for more than 24 hours.
7. Any other injury –
   (a) leading to hypothermia, heat-induced illness or unconsciousness;
   (b) requiring resuscitation; or
   (c) requiring admittance to hospital for more than 24 hours.

8. Loss of consciousness caused by asphyxia or by exposure to a harmful substance or biological agent.
9. Either of the following conditions which result from the absorption of any substance by inhalation, ingestion or through the skin –
   (a) acute illness requiring medical treatment; or
   (b) loss of consciousness.
10. Acute illness which requires medical treatment where there is reason to believe that this resulted from exposure to a biological agent or its toxins or infected material.

Scheduled dangerous occurrences

A dangerous occurrence is a major incident as a rule which has the potential for significant damage and potential loss of life and which is listed in Schedule 2 of RIDDOR. Dangerous occurrences are classified under five headings:

1. General, eg incidents involving lifting machinery, pressure systems, overhead electric lines.
2. Dangerous occurrences which are reportable in mines, eg fire or ignition of gas, escape of gas, insecure tip.
3. Dangerous occurrences which are reportable in respect of quarries, eg misfires, movement of slopes or faces.
4. Dangerous occurrences which are reportable in respect of relevant transport systems, eg accidents involving any kind of train, incidents at level crossings.
5. Dangerous occurrences which are reportable in respect of an offshore workplace, eg releases of petroleum hydrocarbon, fire or explosion.

Reportable diseases

These are diseases listed in Schedule 3 of RIDDOR under three classifications:

1. Conditions due to physical agents and the physical demands of work, eg malignant diseases of bones due to ionising radiation, decompression illness.
2. Infections due to biological agents, eg anthrax, brucellosis, leptospirosis.
3. Conditions due to substances, eg poisoning by carbon disulphide, ethylene oxide and methyl bromide.

Secondly, there are those occupational diseases which are 'prescribed' under the Social Security (Industrial Injuries) (Prescribed Diseases) Regulations 1985, plus various amending regulations, for the purpose of affected individuals claiming disablement benefit. Typical prescribed diseases include lead and manganese poisoning, pneumoconiosis, viral hepatitis and byssinosis. However, as with reportable diseases, the condition is qualified by the particular occupation, eg miner's nystagmus caused by work in or about a mine. (See Chapter 8.)

REPORTING PROCEDURES UNDER THE REGULATIONS

Fatal and major injury accidents, together with scheduled dangerous occurrences, must be:

(a) reported immediately to the enforcing authority by the quickest practicable means, ie telephone, and
(b) reported in writing within 10 days to the enforcing authority on Form 2508.

In the case of an over-three-day injury to a person at work, a written report must be sent to the enforcing authority within seven days of the accident on Form 2508. Occupational diseases listed in Schedule 2 of the regulations must be reported in the same way following receipt of a written report from a doctor, for example, by medical certificate (statutory sick pay form) on Form 2508A. Similarly, gas incidents must be reported on Form 2508G.

In all cases, suitable records must be maintained of such reports. This is best done through retaining photocopies of Forms 2508, 2508A and 2508G in a separate file or register.

ACCIDENT INVESTIGATION

There are two principal reasons for investigating all accidents. Firstly, it is essential to ascertain the cause or causes of the accident, which could be associated with poor standards of machinery safety or an unsafe system of work. Secondly, once the causes are identified, it is essential to prevent a recurrence of that accident. Accidents, particularly fatal accidents, can have a serious effect on the morale of the workforce. They may result in lost production, and damage to structural items, plant and machinery, raw materials and finished products, all of which can cause financial losses to the organisation. In particular, there may be a breach of statute, such as the HASAWA, which could result in prosecution of the company by an inspector of the HSE or local authority. Finally, there may be a need for immediate or planned remedial action with a view to securing legal compliance or to prevent further accidents of the same type. In the majority of cases it will be necessary to undertake investigation of an accident to ensure accurate completion of Form 2508 in the case of reportable accidents. However, there is much to be learnt from selective investigation of minor injury accidents as these frequently result in some form of loss to the organisation. (See 'The costs of accidents' on page 90.)

While it may not be practicable to investigate all accidents, there are a number of aspects which need consideration prior to an investigation, for instance:

(a) the type of accident, eg machinery, handling goods, use of dangerous substances;
(b) the relative severity of injury, eg fractured leg, minor abrasion;
(c) whether the accident is one of a series of similar accidents, eg people slipping on a particular floor surface, thus identifying a trend in accidents;
(d) whether the accident involved machinery, plant and equipment used at work, or certain substances which could be dangerous, eg strong acids;
(e) the possibility of a breach of the law with subsequent need for a clear statement as to the sequence of events leading to the accident, identification of witnesses, taking of statements from such witnesses, and briefing of defending solicitor in the event of pending prosecution;
(f) the accident, no matter how trivial it may seem, could be the subject of a claim by the injured person on the company's employers' liability insurance arrangements; and
(g) the accident could result in a specific course of action by the trade union representing the injured person.

It is important that management is seen to be acting swiftly, particularly in the event of a fatal or major injury accident. Thorough investigation of an accident can result in a number of management actions, for instance:

(a) the identification of specific training needs for certain groups;
(b) detailed analysis of the task through job safety analysis in order to identify the hazards and precautions necessary, together with certain training requirements;
(c) the issue of specific instructions with regard to systems of work, the use of personal protective equipment, or to improve supervision;
(d) increased employee involvement in health and safety activities, for instance, the establishment of a health and safety committee;
(e) the preparation and issue of a company code of practice or guidance note dealing with a particular safety procedure;
(f) establishment of a joint management/worker committee or working party to examine a particular trend in accidents and to report back with recommendations;
(g) establishment of a need for better environmental control, eg improved lighting in a specific area or at a particular job;
(h) identification of the need for improved information with regard to the use and storage of potentially dangerous substances; and
(i) clarification as to the particular responsibilities of senior and line management with regard to health and safety generally.

Cases of occupational disease, sometimes referred to as 'slow accidents' because of the long induction period for some diseases, eg 20 years, also merit careful investigation.

The cause–accident–result sequence (see Figure 3.1) is a useful guide in accident investigation. This sequence shows that there are both direct and indirect causes of accidents but, in many cases, the indirect causes are overlooked in the subsequent investigation. Similarly, there are both direct and indirect results of accidents. The indirect results can be extremely significant, particularly in terms of cost.

'Near misses'

A near miss is an unplanned and unforeseeable incident which could have resulted, but did not result, in death, injury, damage or loss. Studies in the United States by Frank Bird, the exponent of Total Loss Control, showed a correlation between serious or disabling injuries, minor injuries, perhaps requiring first aid treatment only, property damage accidents and accidents with no injury, damage or loss – 'near misses'. These four types of incident were in the ratio 1:10:30:600. In other words, as a result of the studies, for every major or disabling injury, there were 10 minor injuries, 30 property damage accidents and 600 near misses.

What is important is that yesterday's near miss could be tomorrow's fatal accident. Hence the significance of reporting, recording and investigating such incidents, and the implementation of remedial measures to prevent recurrences.

THE COSTS OF ACCIDENTS

All accidents, cases of occupational disease and scheduled dangerous occurrences represent some degree of loss to a company. There are both direct and indirect costs.

Direct costs

These are sometimes referred to as 'insured costs' and involve the company's liabilities both as an occupier of premises and employer of staff. Companies pay premiums to an insurance company to give them cover against claims made by injured persons. Premiums are determined largely by the past claims history and the risks involved in the business operation. Other direct costs are, perhaps, product liability claims for defective or unsafe products or specific injury claims, which may be settled in or out of a court. Fines imposed by courts for breaches of the law, together with defence costs in such cases, can also be substantial direct costs.

INDIRECT CAUSES	DIRECT CAUSES	ACCIDENTS	DIRECT RESULTS	INDIRECT RESULTS
Personal factor *Definition:* Any condition or characteristic of a man that causes or influences him to act unsafely. 1. Knowledge and skill deficiencies: (a) Lack of hazard awareness (b) Lack of job knowledge (c) Lack of job skill. 2. Conflicting motivations: (a) Saving time and effort (b) Avoiding discomfort (c) Attracting attention (d) Asserting independence (e) Seeking group approval (f) Expressing resentment. 3. Physical and mental incapacities.	**Unsafe act** *Definition:* Any act that deviates from a generally recognised safe way of doing a job and increases the likelihood of an accident. **Basic types** 1. Operating without authority 2. Failure to make secure 3. Operating at unsafe speed 4. Failure to warn or signal 5. Nullifying safety devices 6. Using defective equipment 7. Using equipment unsafely 8. Taking unsafe position 9. Repairing or servicing moving or energised equipment 10. Riding hazardous equipment 11. Horseplay 12. Failure to use protection.	**The accident** *Definition:* An unexpected occurrence that interrupts work and usually takes this form of an abrupt contact. **Basic types** 1. Struck by 2. Contact by 3. Struck against 4. Contact with 5. Caught in 6. Caught on 7. Caught between 8. Fall to different level 9. Fall on same level 10. Exposure 11. Overexertion/strain.	**Direct results** *Definition:* The immediate results of an accident. **Basic types** 1. 'No results' or near miss 2. Minor injury 3. Major injury 4. Property damage.	**Indirect results** *Definition:* The consequences for all concerned that flow from the direct result of accidents. **For the injured** 1. Loss of earnings 2. Disrupted family life 3. Disrupted personal life 4. Any other consequences. **For the company** 1. Injury costs 2. Production loss costs 3. Property damage costs 4. Lowered employee morale 5. Poor reputation 6. Poor customer relations 7. Lost supervisor time 8. Product damage costs.
Source causes *Definition:* Any circumstances that may cause or contribute to the development of an unsafe condition. **Major sources** 1. Production employees 2. Maintenance employees 3. Design and engineering 4. Purchasing practices 5. Normal wear through use 6. Abnormal wear and tear 7. Lack of preventive maintenance 8. Outside contractors.	**Unsafe conditions** *Definition:* Any environmental condition that may cause or contribute to an accident. **Basic types** 1. Inadequate guards and safety devices 2. Inadequate warning systems 3. Fire and explosion hazards 4. Unexpected movement hazards 5. Poor housekeeping 6. Protruding hazards 7. Congestion, close clearance 8. Hazardous atmospheric conditions 9. Hazardous placement or storage 10. Unsafe equipment defects 11. Inadequate illumination, noise 12. Hazardous personal attire.			

Figure 3.1 The cause–accident–result sequence

Indirect costs

While many organisations may be fully aware of the direct costs of accidents, very little attention is paid to the indirect costs. Many of these costs may be hidden in other costs and thus not fully recognised, eg production costs, administration costs. Typical indirect costs, many of which can be simply calculated, include the following: treatment costs of the injured employee, eg first aid, transport to hospital, hospital charges, attendance by a local doctor or specialist treatment following the accident; lost time costs, of the injured person, management, first aid staff and others involved; production costs, eg lost production; extra overtime costs to make up production losses; damage costs; and training and supervision costs. It can also be extremely costly to investigate an accident thoroughly in terms of time involvement of management, supervisory staff and witnesses. Other miscellaneous costs include, perhaps, replacement of damaged personal property and incidental costs incurred by witnesses attending court.

There are no current average costs of accidents. A minor injury accident, perhaps requiring 10 minutes for first aid treatment, may cost very little. On the other hand, if such treatments are frequent, costs can soon mount up. In the case of fatal and major injury accidents, both direct and indirect costs can be substantial, frequently going into six figures.

Accident investigation and costing should, therefore, be undertaken by all organisations. Not only does such an exercise identify causes and costs, but it clearly identifies areas of loss and future loss potential, together with providing feedback on future accident prevention strategies. A sample accident costing form is shown in Figure 3.2 on page 93.

SUMMARY

1. There is a legal duty on owners, occupiers and employers to report certain defined injuries, diseases and dangerous occurrences to the enforcement agencies, namely the HSE or local authority.
2. The investigation of accidents should result in a clear indication of the causes of the accident and of the action necessary to prevent a recurrence.
3. All accidents have both direct and indirect costs to an employer.
4. The cause–accident–result sequence provides a useful guide to accident investigation and potential losses resulting from same.

ACCIDENT COSTS ASSESSMENT

Date: Time: Place of accident:

Details of accident:

Injured person
Name in full:
Address:
Occupation:
Injury details:
Length of service:

Accident costs		**£**	**p**
Direct costs			
1. % occupier's liability premium			
2. % increased premiums payable			
3. Claims			
4. Fines and damages awarded in court			
5. Court and legal representation cost			
Indirect costs			
6. *Treatment*	First aid		
	Transport		
	Hospital		
	Others		
7. *Lost time*	Injured person		
	Management		
	Supervisor's		
	First aider's		
	Others		
8. *Production*	Lost production		
	Overtime payments		
	Damage to plant, vehicles, etc		
	Training/supervision replacement		
	Labour		
9. *Investigation*	Management		
	Safety adviser		
	Others, eg safety representatives		
	Liaison with enforcement authority		
10. *Other costs*	Replacement of personal items		
	– injured person		
	– others		
	Other miscellaneous costs		
	Total Costs		

Figure 3.2 A sample accident costing form

REFERENCES

Bird, F E (1974) *Management Guide to Loss Control*, Institute Press, Atlanta, Georgia
Health and Safety Executive (1996) *Reporting a Case of Disease*, HMSO, London
Health and Safety Executive (1996) *Reporting an Injury or a Dangerous Occurrence*, HMSO, London
Morgan, P and Davies, N (1981) The cost of occupational accidents and diseases in Great Britain, *Employment Gazette*, HMSO, London
Reporting of Injuries, Diseases and Dangerous Occurrences Regulations 1995 (SI 1985 No 2023), HMSO, London

4

Principles of Accident Prevention

How often do we hear someone say 'It was an accident! It couldn't be helped'? Others will take the view that an accident is an act of God over which we have no control. Generally, as far as the accident victim is concerned, accidents have the following characteristics. They are unforeseeable, unintended, unexpected and unplanned.

ACCIDENT DEFINITIONS

A number of definitions of the term 'accident' have been put forward over the years, indicating the differing perceptions and views that exist. The following are some examples.

1. Fenton v Thorley & Co Ltd (1903) AC 443

Some concrete happening which intervenes or obtrudes itself upon the normal course of employment. It has the ordinary everyday meaning of an unlooked-for mishap or an untoward event which is not expected or designed by the victim.

2. Royal Society for the Prevention of Accidents (RoSPA)

An unplanned and uncontrolled event which has led to or could have caused injury to persons, damage to plant or other loss.

3. American Institute of Loss Control

An unintended or unplanned happening that may or may not result in personal injury, property damage, work process stoppage or interference, or any combination of these conditions, under such circumstances that personal injury might have resulted.

4. Little Oxford Dictionary

An event without apparent cause; an unexpected event; an unintentional act; a mishap.

5. Health and Safety Unit, University of Aston in Birmingham

An unexpected, unplanned event in a sequence of events that occurs through a combination of causes. It results in physical harm (injury or disease) to an individual, damage to property, a near miss, a loss, business interruption or any combination of these effects.

It is easy to conclude from these definitions that, in addition to their resulting in many cases from a breach of the law (eg failure to guard machinery properly), most accidents represent some form of loss to an organisation. That loss can be quantified in financial terms, eg increased employer's liability premiums, fines in the courts, business interruption types of loss, damage costs, sickness absence costs and production losses.

INCIDENTS

An incident is a form of accident or situation that does not necessarily result in injury to people. The term 'incident' can be defined as 'an undesired event that could (or does) result in loss', or 'an undesired event that could (or does) downgrade the efficiency of the business operation'. Typical incidents could include, for instance, a minor fire, flooding of a warehouse or an adverse product liability incident involving a company's products, which may result in those products being recalled at very short notice. This has been seen in the past with vehicles, food products and certain electrical appliances. Such incidents result in adverse publicity, recall costs and substantial costs in restoring public confidence in that product.

Certain incidents are classified as 'dangerous occurrences' under RIDDOR, as outlined in Chapter 3.

'HAZARD', 'DANGER' AND 'RISK'

Any strategy in preventing accidents at work must draw a distinction between 'hazard', 'danger' and 'risk'. A 'hazard' is defined as 'the result of a departure from the normal situation, which has the potential to cause death, injury, damage or loss'. 'Danger' is defined as 'liability or exposure to harm; a thing that causes peril'. 'Risk', on the other hand, has a number of definitions – 'a chance of bad consequences', 'exposure to mischance', 'exposure to chance of injury or loss', 'the probability of harm, damage or injury' or 'the probability of a hazard leading to personal injury and the severity of that injury'.

Accidents, therefore, are concerned with two specific aspects – namely the actual danger that exists at a particular point in time and, secondly, how people perceive and measure risk. Typical examples of dangerous situations could include an unfenced floor opening, a badly guarded machine, a slippery floor or, in driving situations, snow on the road. However, it is not until people come on to the scene that an accident can take place. Everyone takes risks in varying degrees, and these may be associated with such factors as individual human traits, training and upbringing. The people-related causes of accidents can, therefore, include over-confidence, ignorance, carelessness, lack of training, apathy and inappropriate attitudes to danger and the risks that can be encountered. The philosophy that 'It couldn't happen to me!' is encountered among many people. (See Chapter 5, Human Factors.)

ACCIDENT PREVENTION STRATEGIES

Strategies aimed at preventing accidents should be geared, firstly, to reducing the objective danger in the workplace ('safe place' strategies) and, secondly, to increasing people's perception of the risks at work ('safe person' strategies). The twin concepts of 'safe place' and 'safe person' must, therefore, feature prominently in any company accident prevention programme with the principal objectives of ensuring compliance with the duties laid on persons at work under HASAWA.

A safe workplace

'Safe place' strategies are principally concerned with reducing or eliminating the objective dangers which threaten the safety of workers. These include:

1. *Premises:* Premises should be structurally safe in terms of stability, the soundness of floors, staircases and general means of access and egress. The duty to provide safe premises is further reinforced where members of the public may be invited into a premises.

2. *Environment:* Poor standards of working environment are a contributory factor in many accidents. The provision of a sound working environment implies adequate levels of lighting and ventilation, temperature control and the prevention of environmental stressors, such as noise, vibration, dust and fume emission, all of which can affect the heath of staff.

3. *Plant and machinery:* Legislation, such as the PUWER (see Chapter 11 for further details), requires plant and machinery to be adequately fenced or controlled in such a way that operators are not exposed to risk of injury. All new machinery and plant should be assessed for hazards prior to acquisition. Maintenance and cleaning systems should take into account the safety requirements of staff engaged in such operations.

4. *Materials:* Materials and substances used at work may be toxic or carcinogenic. They may emit harmful dust or fumes during processing, or represent risks of bodily injury during handling. The duties of manufacturers and suppliers of substances used at work are clearly identified in the HASAWA (Section 6), as amended by the CPA 1987, and the CHIP Regulations.

5. *Processes:* A work process may incorporate a number of machines, materials and differing operating skills. Such factors must be considered during process design and be subject to regular monitoring, particularly with regard to the loading and unloading of machines, the use of potentially dangerous substances in processes, the presence of injurious by-products of manufacture, and the levels of skill and supervision necessary.

6. *Systems of work:* The need for clearly defined and documented safe systems of work is abundantly clear in many work situations. The failure to provide such safe systems, to train operators in their use, to supervise and control those systems and to revise them where necessary, are among the principal causes of industrial accidents. Section 2 of the HASAWA clearly identifies the provision of safe systems of work as an employer's duty.

7. *Supervision and control:* Good standards of safety supervision from the boardroom downwards should be indicated in the employer's statement of health and safety policy. The specific health and safety related duties of all levels of management and workers should also be clearly identified in job descriptions.

8. *Training:* It is the legal duty of employers to provide information, instruction, training and supervision under the HASAWA. Attention to safety requirements should be stressed during induction training, on-the-job

training, and training in specific tasks and operations, such as the operation of lifting equipment and the use of permit to work systems.

Safe persons at work

'Safe person' strategies include the following:

1. *Personal protective equipment:* Under the Personal Protective Equipment at Work Regulations 1992, 'personal protective equipment' (PPE) is defined as 'all equipment (including clothing affording protection against the weather) which is intended to be worn or held by a person at work and which protects him against one or more risks to his health or safety, and any addition or accessory designed to meet that objective'.

Employers must ensure that 'suitable' PPE is provided to their employees who may be exposed to a risk to their health or safety while at work, except where and to the extent that the risk has been adequately controlled by other means that are equally or more effective.

PPE shall not be suitable unless:

(a) it is appropriate for the risk or risks involved, the conditions at the place where exposure to the risk may occur, and the period for which it is worn;
(b) it takes account of ergonomic requirements and the state of health of the person or persons who may wear it, and of the characteristics of the workstation of each such person;
(c) it is capable of fitting the wearer correctly, if necessary, after adjustments within the range for which it is designed;
(d) so far as is practicable, it is effective to prevent or adequately control the risk or risks involved without increasing overall risk; and
(e) it complies with any enactment (whether in an Act or instrument) which implements in Great Britain any provision on design or manufacture with respect to health or safety in any of the relevant Community Directives listed in Schedule 1 which is applicable to that item of PPE.

Where it is necessary to ensure that PPE is hygienic and otherwise free of risk to health, every employer and every self-employed person shall ensure that PPE provided under this regulation is provided to a person for use only by him.

Under regulation 6, employers must undertake assessments of PPE prior to choosing the same. Factors for consideration in a PPE assessment are:

(a) an assessment of the risk or risks to health or safety that have not been avoided by other means;

(b) the definition of the characteristics which PPE must have in order to be effective against the risks, taking into account any risks which the equipment itself may create;
(c) comparison of the characteristics of the PPE available with the characteristics referred to in (b) above; and
(d) an assessment as to whether the PPE is compatible with other PPE that is in use and that an employee would be required to wear simultaneously.

Employees must report to their employer any loss of, or obvious defect in, PPE. The provision and use of all forms of personal protective equipment should only be considered as a last resort, ie when all other accident prevention strategies have failed or, alternatively, as an interim measure until one of the 'safe place' strategies mentioned above can be implemented. As an accident prevention strategy, the use of protective equipment relies heavily on the worker wearing the item of personal protection, eg gloves, safety helmet, etc all the time that he may be exposed to the hazard. Factors such as comfort, choice of the equipment involved, ease of movement, ease of putting the item on and removing it, specific job restrictions created by the equipment, the effects of high temperatures, and ease of cleaning, replacement (as with respirators) and maintenance of parts, are significant when considering the use of personal protective equipment as a means of protecting workers from hazards. It should be borne in mind that a high level of supervision and control is necessary to ensure constant use of this equipment.

2. *'Vulnerable' groups:* Certain groups of workers, by virtue of their age, physical condition, lack of experience or even general attitude to work, may be more vulnerable to accidents than others. Such groups include: young persons, whose experience of hazards could be limited; pregnant women, where there may be specific risks to the unborn foetus; disabled persons, whose physical capacity to undertake certain jobs is considerably reduced; and that very small group of 'accident repeaters', who have the same type of accident regularly. Special consideration is needed in these cases.

3. *'Unsafe behaviour':* Horseplay and other forms of unsafe behaviour can be a feature of some work situations if supervision and control are poor. This has particularly been the case with apprentices in the past. Management must take a very strong line here, with instant dismissal of offenders in extreme cases.

4. *Personal hygiene:* Many substances used in industrial processes can promote occupational skin conditions, in particular, dermatitis. Such substances must be properly controlled, and facilities for maintaining good standards of personal hygiene provided and maintained. This includes washing facilities,

which cover showers with hot and cold water, soap, nailbrushes and adequate drying facilities.

5. *Maintaining awareness:* Everyone should be aware of the risks in the workplace. Hazards should be clearly identified in the statement of health and safety policy, together with the precautions necessary on the part of workers. Methods of increasing and maintaining awareness include the use of posters, training, safety competitions, various forms of safety monitoring, hazard-spotting exercises, hazard-reporting systems and the use of 'Days Lost' notice boards indicating the number of days lost per month as a result of accidents at work.

Means of preventing accidents

Strategies for preventing accidents take many forms. These include:

1. *Prohibition:* Some processes and practices may be so inherently dangerous that the only way to prevent accidents is by management placing a total prohibition on that activity. This may take the form of a prohibition on the use of a particular substance, such as an identified toxic substance, or of prohibiting people from carrying out unsafe practices, such as riding on the tines of a fork-lift truck, climbing over moving conveyors or working on roofs without crawlboards.

2. *Substitution:* The substitution of a less dangerous material or system of work will, in many cases, reduce accident potential. Typical examples are the introduction of remote control handling facilities for direct manual handling operations, the substitution of toluene, a much safer substance, for benzene, and the use of non-asbestos substitutes for boiler and pipe lagging.

3. *Change of process:* Design or process engineering can usually change a process to ensure better operator protection. Safety aspects of new systems should be considered in the early stages of projects.

4. *Process control:* This can be achieved through the isolation of a particular process, the use of 'permit to work' systems (see page 103), mechanical or remote control handling systems, restriction of certain operations to highly trained and competent operators, and the introduction of dust and fume arrestment plant.

5. *Safe systems of work:* Formally designated safe systems of work, with high levels of training, supervision and control, are an important strategy in accident prevention (see below).

6. *Personal protective equipment:* This entails the provision of items such as safety boots, goggles, aprons, gloves, etc, but is limited in its application as a safety strategy (as discussed on page 99).

SAFE SYSTEMS OF WORK

A safe system of work is defined as 'the integration of men, machinery and materials in the correct environment to provide the safest possible working conditions in a particular working area'.

A safe system of work should incorporate the following features:

(a) a correct sequence of operations;
(b) a safe working area layout;
(c) a controlled environment in terms of temperature, lighting, ventilation, dust control, humidity control, sound pressure levels and radiation hazards; and
(d) clear specification of safe practices and procedures for the task in question.

Safe systems of work are generally designed through the technique of 'job safety analysis'.

Job safety analysis

Job safety analysis is based on task analysis and takes place in two stages. The first stage entails an examination of the task with particular attention being paid to the following points: the job title; the task operations; the details of machinery, equipment, materials and substances used; the hazards; the degree of risk, ie low, medium or high risk; the specific tasks; the work organisation; and the form of operator protection necessary. Each of these points is detailed on a standard form. The second stage of the operation assesses:

(a) the specific job operations;
(b) the hazards associated with these operations, eg risk of hand injury;
(c) the skills required to perform the task safely in terms of knowledge and individual behaviour;
(d) the external influences on behaviour in terms of:
    (i) the nature of the influence, eg low temperatures;
    (ii) the source of the influence, eg refrigeration system; and
    (iii) the activities involved, eg stacking of refrigerated food products;
(e) the learning method, eg induction training in the use of low temperature protective wear, safe stacking, and the use of the alarm system.

Many organisations now produce safe systems of work on a flowchart basis, copies of which are displayed in the operational area and used in staff training activities.

Permit to work systems

Such a system is a form of safe system of work. It is operated fundamentally where there is a high degree of foreseeable risk, eg entry into confined spaces or vessels containing agitators. A typical permit to work system incorporates a number of clearly defined stages:

(a) Assessment: This stage entails an assessment of the work to be done, the method, materials and equipment to be used, and the inherent hazards. Such an assessment must be made by a senior member of the management team with a view to identifying the safest possible ways to undertake a potentially dangerous task or operation.

(b) Withdrawal from service: The next stage entails withdrawal of the plant from service, designation of this fact by warning signs or fencing around the area, limitation of access in certain cases and restriction of entry to identified personnel.

(c) Isolation: Physical, electrical and/or mechanical isolation of the plant implies controlled restriction of access and the physical locking off of sources of power. In certain cases it may be necessary to undertake environmental testing to ascertain whether the use of breathing apparatus is necessary before entry into a confined space.

(d) Completion of work and return to service: On completion of the scheduled operation, details of which must be clearly identified in the permit to work, the certificate is cancelled and returned to the originator for checking to ensure work has been satisfactorily completed. The plant is then handed back for normal use to the manager responsible for same who, in turn, should check the work has been completed satisfactorily.

All the above headings feature in the permit to work form and, most importantly, require a signature certifying completion of the particular stage before the next stage can proceed.

Permits to work are generally produced in triplicate, with the original going to the person undertaking the work, the first copy to the departmental manager in whose area the work is being undertaken and the second copy being retained by the originator. At completion of the operation and final clearance of the permit, all copies are returned to the originator. It is standard practice to maintain a record of all permits to work issued.

SAFETY MONITORING SYSTEMS

Safety monitoring is concerned with the measurement and evaluation of safety performance. It may take the following forms:

1. *Safety surveys:* This is a detailed examination of a number of critical areas of operation or, perhaps, an in-depth study of all health and safety related activities in a workplace.

2. *Safety tours:* These are an unscheduled examination of a working area, frequently undertaken as a group exercise (eg foreman, safety representative and safety committee member), to assess general compliance with safety requirements (eg fire protection measures and use of machinery safety devices).

3. *Safety audits:* A safety audit fundamentally subjects each area of an organisation's activities to a systematic critical examination with the object of minimising injury and loss. It generally takes the form of a series of questions directed to examining factors such as the operation of safe systems of work, compliance with the statement of health and safety policy and the operation of hazard reporting systems. A specimen audit form can be seen on pages 105–109.

4. *Safety inspections:* A scheduled inspection of a premises or working area to assess levels of legal compliance and observation of company safety procedures. Safety inspections are frequently undertaken by company safety specialists and trade union safety representatives.

5. *Safety sampling:* A system designed to measure by random sampling the accident potential in a workplace or process by identifying defects in safety performance or omissions. Observers follow a prescribed route through the working area noting deficiencies in performance, eg concerning the wearing of personal protective equipment or the use of correct manual handling techniques. In some cases, individual topics in the safety sampling exercise are ranked according to importance with a maximum number of points achievable. At the end of the exercise a total score is identified which gives an indication of the performance level at that point in time. A typical safety sampling exercise is shown in Figure 4.1, Health and safety review.

6. *Hazard and operability studies:* Such studies incorporate the application of formal critical examination to the process and engineering intentions regarding new facilities. The principal aim of such a study is to assess the

hazard potential arising from the incorrect operation of equipment and the consequential effects on the facility. Such an operation enables remedial action to be taken at a very early stage.

7. *Damage control:* Levels of damage are an indication of future accident potential. Damage control operates on the philosophy that non-injury accidents are just as important as injury accidents. The elimination of the causes of accidents resulting in damage to property, plant and products frequently results in a reduction in injury accidents.

Specimen Safety Audit

	YES	NO
1. DOCUMENTATION		
1. Are management aware of all health and safety legislation applying to their workplace? Is this legislation available to management and employees?		
2. Have all Approved Codes of Practice, HSE Guidance Notes and internal codes of practice been studied by management with a view to ensuring compliance?		
3. Does the existing statement of health and safety policy meet current conditions in the workplace? Is there a named manager with overall responsibility for health and safety? Are the 'organisation and arrangements' to implement the health and safety policy still adequate? Have the hazards and precautions necessary on the part of staff and other persons been identified and recorded? Are individual responsibilities for health and safety clearly detailed in the statement?		
4. Do all job descriptions adequately describe individual health and safety responsibilities and accountabilities?		
5. Do written safety guidelines exist for all potentially hazardous operations? Is permit to work documentation available?		
6. Has a suitable and sufficient assessment of the risks to staff and other persons been made, recorded and brought to the attention of staff and other persons? Have other risk assessments in respect of: (a) substances hazardous to health; (b) risks to hearing; (c) work equipment; (d) personal protective equipment; (e) manual handling operations; (f) display screen equipment; and (g) fire, been made, recorded and brought to the attention of staff and other persons?		

	YES	NO
7. Is there a record of inspections of the means of escape in the event of fire, fire appliances, fire alarms, warning notices, fire and smoke detection equipment?		
8. Is there a record of inspections and maintenance of work equipment, including guards and safety devices? Are all examination and test certificates available, eg lifting appliances and pressure systems?		
9. Are all necessary licences available, eg to store petroleum spirit?		
10. Are workplace health and safety rules and procedures available, promoted and enforced? Have these rules and procedures been documented in a way which is comprehensible to staff and others, eg health and safety handbook? Are disciplinary procedures for unsafe behaviour clearly documented and known to staff and other persons?		
11. Is a formally written emergency procedure available?		
12. Is documentation available for the recording of injuries, near misses, damage only accidents, diseases and dangerous occurrences?		
13. Are health and safety training records maintained?		
14. Are there documented procedures for regulating the activities of contractors, visitors and other persons working on the site?		
15. Is hazard reporting documentation available to staff and other persons?		
16. Is there a documented planned maintenance system?		
17. Are there written cleaning schedules?		
2. HEALTH AND SAFETY SYSTEMS		
1. Have competent persons been appointed to: (a) co-ordinate health and safety measures; and (b) implement the emergency procedure? Have these persons been adequately trained on the basis of identified and assessed risks? Is the role, function, responsibility and accountability of competent persons clearly identified?		
2. Are there arrangements for specific forms of safety monitoring, eg safety inspections, safety sampling? Is a system in operation for measuring and monitoring individual management performance on health and safety issues?		
3. Are systems established for the formal investigation of accidents, ill health, near misses and dangerous occurrences? Do investigation procedures produce results which can be used to prevent future incidents? Are the causes of accidents, ill health, near misses and dangerous occurrences analysed in terms of the failure of established safe systems of work?		
4. Is a hazard reporting system in operation?		
5. Is a system for controlling damage to structural items, machinery, vehicles, etc in operation?		
6. Is the system for joint consultation with trade union safety representatives and staff effective?		

YES	NO

Are the role, constitution and objectives of the health and safety committee clearly identified?
Are the procedures for appointing or electing committee members and trade union safety representatives clearly identified?
Are the available facilities, including training arrangements, known to committee members and trade union safety representatives?

7. Are the capabilities of employees as regards health and safety taken into account when entrusting them with tasks?
8. Is the provision of first aid arrangements adequate?
   Are first aid personnel adequately trained and retrained?
9. Are the procedures covering sickness absence known to staff?
   Is there a procedure for controlling sickness absence?
   Are managers aware of the current sickness absence rate?
10. Do current arrangements ensure that health and safety implications are considered at the design stage of projects?
11. Is there a formal annual health and safety budget?

3. PREVENTION AND CONTROL PROCEDURES

1. Are formal inspections of machinery, plant, hand tools, access equipment, electrical equipment, storage equipment, warning systems, first aid boxes, resuscitation equipment, welfare amenity areas, etc undertaken?
   Are machinery guards and safety devices examined on a regular basis?
2. Is a permit to work system operated where there is a high degree of foreseeable risk?
3. Are fire and emergency procedures practised on a regular basis?
   Where specific fire hazards have been identified, are they catered for in the current fire protection arrangements?
   Are all items of fire protection equipment and alarms tested, examined and maintained on a regular basis?
   Are all fire exits and escape routes marked, kept free from obstruction and operational?
   Are all fire appliances correctly labelled, sited and maintained?
4. Is a planned maintenance system in operation?
5. Are the requirements of cleaning schedules monitored?
   Is housekeeping of a high standard, eg material storage, waste disposal, removal of spillages?
   Are all gangways, stairways, fire exits, access and egress points to the workplace maintained and kept clear?
6. Is environmental monitoring of temperature, lighting, ventilation, humidity, radiation, noise and vibration undertaken on a regular basis?
7. Is health surveillance of persons exposed to assessed health risks undertaken on a regular basis?
8. Is monitoring of personal exposure to assessed health risks undertaken on a regular basis?
9. Are local exhaust ventilation systems examined, tested and maintained on a regular basis?
10. Are arrangements for the storage and handling of substances hazardous to health adequate?

	YES	NO
Are all substances hazardous to health identified and correctly labelled, including transfer containers?		
11. Is the appropriate personal protective equipment available? Is the personal protective equipment worn or used by staff consistently when exposed to risks? Are storage facilities for items of personal protective equipment provided?		
12. Are welfare amenity provisions, ie sanitation, hand washing, showers and clothing storage arrangements adequate? Do welfare amenity provisions promote appropriate levels of personal hygiene?		
4. INFORMATION, INSTRUCTION, TRAINING AND SUPERVISION		
1. Is the information provided by manufacturers and suppliers of articles and substances for use at work adequate? Do employees and other persons have access to this information?		
2. Is the means of promoting health and safety adequate? Is effective use made of safety propaganda, eg posters?		
3. Do safety signs meet the requirements of the Safety Signs Regulations 1980 and Health and Safety (Safety Signs and Signals) Regulations 1996? Are safety signs adequate in terms of the assessed risks?		
4. Are fire instructions prominently displayed?		
5. Are hazard warning systems adequate?		
6. Are the individual training needs of staff and other persons assessed on a regular basis?		
7. Is staff health and safety training undertaken:		
(a) at the induction stage;		
(b) on employees being exposed to new or increased risks because of:		
(i) transfer or change in responsibilities;		
(ii) the introduction of new work equipment or a change respecting existing work equipment;		
(iii) the introduction of new technology;		
(iv) the introduction of a new system of work or change in an existing system of work?		
Is the above training:		
(a) repeated periodically;		
(b) adapted to take account of new or changed risks; and		
(c) carried out during working hours?		
8. Is specific training carried out regularly for first aid staff, fork-lift truck drivers, crane drivers and others exposed to specific risks? Are selected staff trained in the correct use of fire appliances?		
5. FINAL QUESTION		
Are you satisfied that your organisation is as safe and healthy as you can reasonably make it, or that you know what action must be taken to achieve that state?		

ACTION PLAN

1. Immediate action

2. Short-term action (14 days)

3. Medium-term action (6 months)

4. Long-term action (2 years)

_________________________Auditor Date________________

ACCIDENT AND ILL-HEALTH RATES

The following rates are used in the calculation of accident and ill-health rates:

$$\text{Frequency Rate} = \frac{\text{Total number of accidents}}{\text{Total number of man-hours worked}} \times 100{,}000$$

$$\text{Incident Rate} = \frac{\text{Total number of accidents}}{\text{Number of persons employed}} \times 1{,}000$$

$$\text{Severity Rate} = \frac{\text{Total number of days lost}}{\text{Total number of man-hours worked}} \times 1{,}000$$

Unit .... Date ....

WORKPLACE	MAX	RATING	SAFETY SYSTEMS	MAX	RATING
Cleaning & housekeeping	10		Policy statement & plan	10	
Structural safety	10		Safe systems of work	20	
Machinery safety	20		Accident reporting/costing	10	
Fire protection	20		Sickness absence reporting/ costing	10	
Chemical safety	10		Hazard reporting	10	
Electrical safety	10		Cleaning schedules	10	
Internal transport	10		Catering hygiene & safety	10	
Access equipment	5		Emergency procedure	10	
Internal storage	5		Safety monitoring procedure	10	
MAX	100		MAX	100	
PEOPLE			ENVIRONMENT		
Personal protection	20		Environmental control – internal	10	
Chemical handling	10		Environmental control – external	5	
Manual handling	10		Noise control – internal	10	
Propaganda	20		Noise control – external	5	
Training	20		Welfare amenity provisions	20	
Safe behaviour	20		MAX	50	
MAX	100		TOTAL SCORE MAX	350	

Figure 4.1 Health and safety review

$$\text{Mean Duration Rate} = \frac{\text{Total number of days lost}}{\text{Total number of accidents}}$$

$$\text{Duration Rate} = \frac{\text{Number of man-hours worked}}{\text{Total number of accidents}}$$

$$\text{Sickness Absence Rate} = \frac{\text{Total man-days lost through sickness}}{\text{Total man-days worked}} \times 100$$

RISK ASSESSMENT

A risk assessment may be defined as: 'an identification of the hazards present in an undertaking and an estimate of the extent of the risks involved, taking into account whatever precautions are already being taken'.

It is essentially a four-stage process:

(a) identification of all the hazards;
(b) measurement of the risks;
(c) evaluation of the risks; and
(d) implementation of measures to eliminate or control the risks.

There are different approaches which can be adopted in the workplace, eg:

(a) examination of each activity which could cause injury;
(b) examination of hazards and risks in groups, eg machinery, substances, transport; and / or
(c) examination of specific departments, sections, offices, construction sites.

In order to be suitable and sufficient and to comply with legal requirements, a risk assessment must:

(a) identify all the hazards associated with the operation, and evaluate the risks arising from those hazards, taking into account current legal requirements;
(b) record the significant findings if more than five persons are employed, even if located in different locations;
(c) identify any group of employees, or single employees as the case may be, who are especially at risk;
(d) identify others who may be specially at risk, eg visitors, contractors, members of the public;
(e) evaluate existing controls, stating whether or not they are satisfactory and, if not, what action should be taken;

(f) evaluate the need for information, instruction, training and supervision;
(g) judge and record the likelihood of an accident occurring as a result of uncontrolled risk, including the 'worst case' likely outcome;
(h) record any circumstances arising from the assessment where serious and imminent danger could arise; and
(i) provide an action plan giving information on implementation of additional controls, in order of priority, and with a realistic time-scale.

Recording the assessment

The assessment must be recorded in organisations where more than five persons are employed. (Electronic methods of recording are acceptable.) The assessment must incorporate details of items (a) to (i) above.

Generic assessments

These are assessments produced once only for a given activity or type of workplace. In cases where an organisation has several locations, or in situations where the same activity is undertaken, a generic risk assessment could be carried out for a specific activity to cover all locations. Similarly, where operators work away from the main location and undertake a specific task, eg installation of telephones or servicing of equipment, a generic assessment should be produced.

For generic assessments to be effective:

(a) 'worst case' situations must be considered; and
(b) provision should be made for monitoring implementation of assessment controls which are / are not relevant at a particular location, and for determining what action needs to be taken to implement the relevant tasks outlined in the assessment.

In certain cases, there may be risks which are specific to one situation only, and these risks may need to be incorporated in a separate part of the generic risk assessment.

Maintaining the risk assessment

The risk assessment must be *maintained*. This means that any significant change to a workplace, process or activity, or the introduction of any new process, activity or operation, should be subject to risk assessment. If new hazards come to light, then these should also be subject to risk assessment. The risk assessment, furthermore, should be periodically reviewed and updated.

This is best achieved by a suitable combination of safety inspection and monitoring techniques, which require corrective and/or additional action where the need is identified.

Typical monitoring systems include:

(a) preventive maintenance inspections;
(b) safety representative/committee inspections;
(c) statutory and maintenance scheme inspections, tests and examinations;
(d) safety tours and inspections;
(e) occupational health surveys;
(f) air monitoring; and
(g) safety audits.

Useful information on checking performance against control standards can also be obtained reactively from the following activities:

(a) accident and ill-health investigation;
(b) investigation of damage to plant, equipment and vehicles; and
(c) investigation of 'near miss' situations.

Reviewing the risk assessment

The frequency of review depends upon the level of risk in the operation, and should not normally exceed 10 years. Further, a review is necessary if a serious accident occurs in the organisation or elsewhere; or a check on the risk assessment shows a gap in assessment procedures.

Risk/hazard control

Once the risk or hazard has been identified and assessed, employers must either prevent the risk arising or, alternatively, control same. Much will depend upon the magnitude of the risk in terms of the controls applied. In certain cases, the level of competence of operators may need to be assessed prior to their undertaking certain work, eg work on electrical systems.

A typical hierarchy of control, from high-risk to low-risk, is indicated below:

1. **Elimination** of the risk completely, eg prohibiting a certain practice or the use of a certain hazardous substance.
2. **Substitution** with something less hazardous or risky.
3. **Enclosure** of the risk in such a way that access is denied.
4. **Fitting guards or installing safety devices** to prevent access to danger points or zones on work equipment and machinery.
5. **Safe systems of work** that reduce the risk to an acceptable level.

6. **Written procedures,** eg job safety instructions, that are known and understood by those affected.
7. **Adequate supervision,** particularly in the case of young or inexperienced persons.
8. **Training** of staff to appreciate the risks and hazards.
9. **Information,** eg safety signs, warning notices.
10. **Personal protective equipment,** eg eye, hand, head and other forms of body protection.

In many cases, a combination of the above control methods may be necessary.

It should be appreciated that the amount of management control necessary will increase proportionately for the controls lower down this list. In other words, item 1 indicates no management control is needed, whereas item 10 requires a high degree of control.

COMPETENT PERSONS

In the last decade considerable attention has been paid to the requirement for organisations to appoint 'competent persons'. However, this concept is not new. Case law, namely Brazier *v* Skipton Rock Company Ltd (1962) (1 AER 955) states that 'a competent person should have practical and theoretical knowledge as well as sufficient experience of the particular plant, machinery or procedure involved to enable him to identify defects or weaknesses during plant and machinery examinations, and to assess their importance in relation to the strength and function of that plant and machinery'.

This definition arises from duties particularly under former construction safety law with respect to the requirement for competent persons, for example to inspect scaffold materials prior to erection of a scaffold, to inspect excavations on a daily basis and to supervise the erection of cranes.

Regulation 7 of the MHSWR refers to 'health and safety assistance' and outlines the duties of employers in the appointment of competent persons. (See Chapter 1.) More detailed information is provided in the ACOP to the regulations thus:

> 'Employers are solely responsible for ensuring that those they appoint to assist them with health and safety measures are competent to carry out the tasks they are assigned and are given adequate information and support. In making decisions on who to appoint, employers themselves need to know and understand the work involved, the principles of risk assessment and prevention, and current legislation and health and safety standards. Employers should ensure that anyone they appoint is capable of applying the above to whatever task they are assigned.

> 'Employers must have access to competent help in applying the provisions of health and safety law, including these Regulations. In particular they need competent help in devising and applying protective measures, unless they are competent to undertake the measures without assistance. Appointment of competent people for this purpose should be included among the health and safety arrangements under regulation 5(2). Employers are required by the Safety Representatives and Safety Committees Regulations 1977 to consult safety representatives in good time on arrangements for the appointment of competent assistance.'

The HSE Guidance accompanying the MHSWR makes the following points:

> 'The appointment of health and safety assistants or advisers does not absolve the employer from responsibilities for health and safety under the HASAWA and other relevant statutory provisions. It can only give added assurance that these responsibilities will be discharged adequately. Where external services are employed, they will usually be appointed in an advisory capacity only.
>
> 'Competence in the sense that it is used in these Regulations does not necessarily depend on the possession of particular skills or qualifications. Simple situations may require only the following:

(a) an understanding of relevant current best practice;
(b) an awareness of the limitations of one's own experience and knowledge; and
(c) the willingness and ability to supplement existing experience and knowledge, when necessary by obtaining external help and advice.

> 'More complicated situations will require the competent assistant to have a higher level of knowledge and experience. More complex or highly technical situations will call for specific applied knowledge and skills which can be offered by appropriately qualified specialists. Employers are advised to check the appropriate health and safety qualifications (some of which may be competence-based and/or industry specific), or membership of a professional body or similar organisation (at an appropriate level and in an appropriate part of health and safety) to satisfy themselves that the assistant they appoint has a sufficiently high level of competence. Competence-based qualifications accredited by the Qualifications and Curriculum Authority and the Scottish Qualifications Authority may also provide a guide.'

In addition to the more general duties under the MHSWR, competent persons have specific duties under various regulations. These duties can be stated thus.

Pressure Systems Safety Regulations 2000

Owners and users of pressure systems must have a written scheme of examination drawn up by a competent person for the examination of the system at specified intervals. Competence must be based on the type of work undertaken in the case of a major, intermediate or minor pressure system (as defined).

The competent person:

(a) advises the user on the scope of the written scheme of examination;
(b) draws up or certifies the scheme of examination; and
(c) undertakes examinations under the scheme.

A competent person must be sufficiently independent from the interests of all other functions to ensure adequate segregation of accountabilities.

Electricity at Work Regulations 1989

No person shall carry out a work activity where technical knowledge or experience is necessary to prevent danger or injury unless he has such knowledge or is under the appropriate degree of supervision. (While the term 'competent person' does not appear in the regulations, competence is implied in this case.)

Construction (Design and Management) Regulations 2007

No person on whom these regulations place a duty shall:

(a) appoint or engage a CDM co-ordinator, designer, principal contractor or contractor unless he has taken reasonable steps to ensure that the person to be appointed or engaged is competent;
(b) accept such an appointment or engagement unless he is competent;
(c) arrange for or instruct a worker to carry out or manage design or construction work unless the worker is:
    (i) competent; or
    (ii) under the supervision of a competent person.

Regulatory Reform (Fire Safety) Order 2004

The responsible person must appoint one or more competent persons to assist him in undertaking the preventive and protective measures.

Lifting Operations and Lifting Equipment Regulations 1998

These regulations require that:

(a) lifting operations must be planned, supervised and carried out in a safe manner by a competent person;
(b) equipment used for lifting people must be marked accordingly, and should be safe for such a purpose;
(c) before being used for the first time the equipment should, where appropriate, be 'thoroughly examined';
(d) equipment may also need to be 'thoroughly examined' in use at set intervals, for example six monthly for accessories and equipment used for lifting people;
(e) all examinations must be carried out by a competent person; and
(f) a report must be submitted by the competent person to the employer following a thorough examination or inspection of the equipment.

THE ROLE OF THE HEALTH AND SAFETY PRACTITIONER

Practitioners in occupational health and safety include occupational health nurses, health and safety managers/advisers, safety officers, occupational physicians, occupational hygienists and safety representatives/stewards. Their various roles are outlined below.

Occupational health nurses

Generally an occupational health nurse would be a qualified nurse, eg RGN, with a separate qualification in occupational health nursing, for instance the Occupational Health Nursing Certificate (OHNC).

The occupational health nurse's role consists of eight main elements:

(a) health supervision;
(b) health education;
(c) environmental monitoring and occupational safety;
(d) counselling;
(e) treatment services;
(f) rehabilitation and resettlement;
(g) unit administration and record systems; and
(h) liaison with other agencies, eg employment medical advisers of the Employment Medical Advisory Service.

The Royal College of Nursing lists the following duties which could be undertaken by a fully trained occupational health nurse:

(a) health assessment in relation to the individual worker and the job to be performed;
(b) noting normal standards of health and fitness and any departures or variations from these standards;
(c) referring to the occupational physician or doctor such cases which, in the opinion of the nurse, require further investigation and medical, as distinct from nursing, assessment;
(d) health supervision of vulnerable groups, eg young persons;
(e) routine visits to and surveys of the working environment, and informing as necessary the appropriate expert when a particular problem requires further specialised investigation;
(f) employee health counselling;
(g) health education activities in relation to groups of workers;
(h) the assessment of injuries or illness occurring at work and treatment or referral as appropriate;
(i) responsibility for the organisation and administration of occupational health services, and the control and safe-keeping of non-statutory personal health records; and
(j) a teaching role in respect of the training of first aid personnel and the organisation of emergency services.

Health and safety managers/advisers and safety officers

The increase in attention to safety at work following the HASAWA has resulted, to some extent, in the development, particularly in large multisite organisations, of a two-tier system of specialists in occupational health and safety. The larger companies tend to promote this two-tier system with a health and safety manager, frequently reporting at director or board level, supported by local safety officers who operate in a specific work location. Health and safety managers / advisers are concerned with advising the organisation on policy issues and with ensuring the integration of health and safety considerations into the normal operating procedures of the organisation. They may come from a number of backgrounds, eg HSE, or from particular disciplines such as chemistry, engineering, etc.

The health and safety manager would formulate policy on matters relating to, for instance, the development of local statements of health and safety policy, health and safety training, promotion of health and safety related issues, control of the working environment, disaster planning and control, and accident reporting procedures. In some cases he may be responsible for monitoring the performance of full-time or part-time safety officers in individual units. His objectives are principally concerned with reducing losses, improving management performance and generally raising the profile of health and safety in the organisation.

Safety officers may be appointed on a full-time or part-time basis. Their role and function are specifically related to the workplace in which they operate. While there is no current legal requirement for such persons to be trained, increasingly companies are seeking membership of the Institution of Occupational Safety and Health (IOSH) as a basic qualification for this task. Training courses at Certificate and Diploma level are organised through the National Examination Board in Occupational Safety and Health (NEBOSH), and such courses can be taken on a full-time, part-time or block-release basis at local technical colleges, through the Royal Society for the Prevention of Accidents (RoSPA) and other approved training providers.

IOSH has indicated the following as typical duties of a safety officer:

(a) advising line management in order to assist it to fulfil its responsibility for safety, including:
    (i) advising on safety aspects in the design and use of plant and equipment and the checking of new equipment before commissioning; and
    (ii) carrying out periodic inspections to identify unsafe plant, unsafe working conditions and unsafe practices, to report on the results of such inspections and to make recommendations for remedying any defects found;
(b) advising upon the drawing up and implementation of safe systems of work, and the provision and use of personal protective equipment;
(c) advising on legal requirements affecting health and safety;
(d) participating in the work of safety committees and joint consultations affecting the workforce;
(e) promoting and, where appropriate, participating in safety education programmes to assist in safety consciousness at all levels within the organisation and specifically to teach supervisors to develop safe working conditions;
(f) working in collaboration with the training department, where this exists, to secure regular safety training of employees;
(g) providing information about accident prevention techniques and preparing visual aids, including posters, slides, film strips, etc for safety training purposes;
(h) assessing possible causes of injury and circumstances likely to produce an accident, the compilation of necessary reports, and tendering advice to prevent recurrence;
(i) recording of accident statistics and presenting information in appropriate form for the use of management and others in ensuring safety performance;
(j) maintaining liaison with other departments, including medical and training departments, with official bodies such as government inspectorates, local

authorities and fire authorities, and with outside bodies such as RoSPA and the Fire Protection Association; and

(k) keeping up to date with modern processes and techniques, with special reference to safety.

Occupational physicians

An occupational physician is a registered medical practitioner. He should preferably hold a recognised qualification in occupational medicine, such as membership of the Faculty of Occupational Medicine or other appropriate organisation. The British Medical Association has identified the role of the occupational physician as encompassing the following:

1. The effect of health on the capacity to work, which includes:

(a) provision of advice to employees on all health matters relating to their working capacity;
(b) examination of applicants for employment and advice as to their placement;
(c) immediate treatment of surgical and medical emergencies occurring at the place of employment;
(d) examination and continued observation of persons returning to work after absence due to illness or accident and advice on suitable work;
(e) health supervision of all employees with special reference to young persons, pregnant women, elderly persons and disabled persons.

2. The effect of work on health, which includes:

(a) responsibility for nursing and first aid services;
(b) study of the work and working environment and how they affect the health of employees;
(c) periodical examination of employees exposed to special hazards in respect of their employment;
(d) advising management regarding:
    (i) the working environment in relation to health;
    (ii) occurrence and significance of hazards;
    (iii) accident prevention; and
    (iv) statutory requirements in relation to health;
(e) medical supervision of the health and hygiene of staff and facilities, with particular reference to canteens, kitchens, etc and those working in the production of foods or drugs for sale to the public;
(f) arranging and carrying out such education work in respect of the health, fitness and hygiene of employees as may be desirable and practicable;
(g) advising those committees within the organisation which are responsible for the health, safety and welfare of employees.

Occupational hygienists

'Occupational hygiene' has been defined as being concerned with the identification, measurement, evaluation and control of environmental factors, such as noise and radiation, and contaminants arising from work operations which may adversely affect the health of employees.

Entry to the profession is controlled by the British Examination Board in Occupational Hygiene (BEBOH). Occupational hygienists frequently operate on a consultancy basis, undertaking measurement and evaluation of airborne hazards, such as asbestos, fumes, dusts and vapours, in addition to physical phenomena, such as noise, vibration and radiation. The significance of occupational hygiene as a specialist role in occupational health and safety has increased with the advent of the COSHH Regulations.

SUMMARY

1. All accidents and incidents represent varying degrees of loss to organisations.
2. Accident prevention strategies should take into account the concepts of 'safe place' and 'safe person', with the emphasis being on 'safe place' strategies.
3. Safe systems of work are the basis for effective levels of safety management.
4. All organisations should operate one or more forms of safety monitoring as a means of establishing standards and measuring performance against these standards.
5. When considering the appointment of health and safety practitioners, reference should be made to the accepted standards of competence and professional qualifications outlined in this chapter.

CONCLUSION TO PART I

Part I has examined the management and administration aspects of occupational health and safety. Of particular significance are the duties of employers and employees under the HASAWA and the importance of clearly written statements of health and safety policy.

The policy statement forms the framework for all health and safety related activities in the organisation. In particular, it identifies the duties and responsibilities of all levels of management and employees in the organisation and arrangements for ensuring implementation of the policy and systems for measuring performance.

However, the legal requirements should be considered the minimum acceptable standard. There should now be a move away from the purely legalistic approach to health and safety towards a philosophy of 'asset protection'. Without people, companies would not be in business; the loyal, skilled, long-serving employee is any company's most valuable asset. The maintenance of good standards of safety, health and welfare is one of the most effective ways of ensuring continuity of that loyalty.

Frank Bird, the American exponent of Total Loss Control, stated many years ago that 'Accidents downgrade the system.' Apart from resulting in substantial losses to a company, they result in reduced morale, poor industrial relations, loss of company image in the market place and loss of public confidence. The need, therefore, for clearly developed safe systems of work, accident reporting procedures and well-developed safety monitoring systems cannot be over-emphasised. Such operations should form part of the company 'culture' and not be viewed in isolation as is so frequently the case.

Health and safety practitioners have a significant contribution to make in reducing the losses. However, they must be properly trained and set clearly defined objectives which are capable of measurement.

REFERENCES

Atherley, G R C (1978) *Occupational Health and Safety Concepts*, Applied Science Publishers Ltd, London

Bird, F E and Loftus, R G (1984) *Loss Control Management*, RoSPA, Birmingham

Chemical Industries Association Ltd (1975) *Safety Audits*, Chemical Industries Association, London

Department of Employment (1974) *Accidents in Factories: The pattern of causation and scope for prevention*, HMSO, London

Health and Safety Executive (1982) *Guidelines for Occupational Health Services: Guidance Note HS(G) 20*, HMSO, London

Health and Safety Executive (1992) *Personal Protective Equipment at Work Regulations 1992 and Guidance on Regulations*, HMSO, London

Health and Safety Commission (1999) *Management of Health and Safety at Work Regulations 1999 and Approved Code of Practice*, HMSO, London

Stranks, J (1995) *Management Systems for Safety*, Pitman, London

Stranks, J (1996) *The Law and Practice of Risk Assessment*, Pitman, London

Stranks, J (2005) *Handbook of Health and Safety Practice*, Pitman, London

Part 2

People at Work

5

Human Factors

The MHSWR emphasised the need for employers to place greater emphasis on people, their capabilities and fallibilities. Regulation 13 states that 'every employer shall, in entrusting tasks to his employees, take into account their capabilities as regards health and safety'.

The ACOP qualifies this requirement as follows: 'When allocating work to employees, employers should ensure that the demands of the job do not exceed the employees' ability to carry out the work without risk to themselves or others. Employers should take account of the employees' capabilities and the level of their training, knowledge and experience. If additional training is needed, it should be provided.'

The HSE publication *Reducing Error and Influencing Behaviour* (HSG 48) defines 'human factors' as a range of issues including the perceptual, physical and mental capabilities of people and the interactions of individuals with their job and the working environments, the influence of equipment and system design on human performance and, above all, the organisational characteristics which influence safety related behaviour at work.

There are three areas of influence on people at work, namely:

(a) the organisation;
(b) the job; and
(c) personal factors.

These are directly affected by:

(a) the system for communication within the organisation and
(b) the training systems and procedures in operation,

all of which are directed at preventing human error.

The organisation

Characteristics of organisations which influence safety-related behaviour include:

(a) the need to promote a positive climate in which health and safety is seen by both management and employees as being fundamental to the organisation's day-to-day operations, ie they must create a positive safety culture;
(b) the need to ensure that policies and systems which are devised for the control of risk from the organisation's operations take proper account of human capabilities and fallibilities;
(c) commitment to the achievement of progressively higher standards which is shown at the top of the organisation and cascaded through successive levels of same;
(d) demonstration by senior management of their active involvement, thereby galvanising managers throughout the organisation into action; and
(e) leadership whereby an environment is created which encourages safe behaviour.

The job

Successful management of human factors and the control of risk involves the development of systems designed to take proper account of human capabilities and fallibilities. Using techniques like job safety analysis, jobs should be designed in accordance with ergonomic principles (see Chapter 6) so as to take into account limitations in human performance.

Major considerations in job design include:

(a) identification and comprehensive analysis of the critical tasks expected of individuals and appraisal of likely errors;
(b) evaluation of required operator decision making and the optimum balance between human and automatic contributions to safety actions;
(c) application of ergonomic principles in the design of man–machine interfaces, including displays of plant and process information, control devices and panel layouts;
(d) design and presentation of procedures and operating instructions;
(e) organisation and control of the working environment including workspace, access for maintenance, noise, lighting and thermal conditions;

(f) provision of correct tools and equipment;
(g) scheduling of work patterns, including shift organisation, control of fatigue and stress, and arrangements for emergency operations; and
(h) efficient communications, both immediate and over periods of time.

Personal factors

This aspect is concerned with how factors such as attitude, motivation, training, human error and the perceptual, physical and mental capabilities of people can interact with health and safety issues.

Attitudes are directly connected with an individual's self-image, the influence of groups and compliance with group norms or standards and, to some extent, opinions, including superstitions, like 'all accidents are acts of God!' Changing attitudes is difficult. They may be affected by past experience, the level of intelligence, specific motivation, financial gain and by the skills available to the individual. There is no doubt, however, that management example is the strongest of all motivators for bringing about attitude change.

Important factors in motivating people to work safely include joint consultation in planning the work organisation, the use of working parties to define objectives, attitudes currently held, the system for communication within the organisation and the quality of leadership at all levels. Research has shown that financially related motivation systems, such as the payment of safety bonuses, do not necessarily change attitudes, people reverting to normal behaviour when the motivator is removed.

BEHAVIOURAL SAFETY

Professor Dominic Cooper, the well-known exponent of behavioural safety, defines the subject as 'the systematic application of psychological research on human behaviour to the problems of safety in the workplace'.

It is the process of involving employees in defining the ways that they are most likely to be injured, seeking their involvement, obtaining their 'buy in', and asking them to observe and monitor co-employees with a view to reducing their unsafe behaviours. Fundamentally, it is a means of obtaining increased improvements in safety performance through the promotion of safe behaviours at all levels in the workplace and in an organisation.

There are many approaches to behavioural safety, but the ultimate objective of this proactive approach to safety improvement is, principally, to predict and provide an early warning of accidents and loss-producing incidents arising from work activities. Most importantly, it is not a replacement for the more traditional proactive and reactive approaches to accident prevention based on safety monitoring procedures, risk assessment and the investigation of accidents, incidents and occupational ill health. However, it does challenge these more traditional approaches which may not necessarily take into account the potential for human error as a contributory factor in accidents.

This approach is directed at improving self-awareness, encouraging contributions from, and the involvement of, people at work in recommending the improvements necessary, the sharing of feedback on safety performance, and increasing people's perception of risk. It is also concerned with establishing and promoting the right safety culture within an organisation, based on senior management commitment and a demonstration of that commitment at all levels within an organisation.

Setting the standards

If behavioural safety is going to work effectively, there must be formally established systems and operating procedures. The establishment of these systems and procedures is an on-going process commencing with those operations and activities involving the greatest danger to employees. Techniques such as job safety analysis and risk assessment are useful tools in identifying the significant hazards and the precautions necessary on the part of employees.

At this standard-setting stage it may be necessary to acquire information from official sources, such as the HSE, RoSPA and the Institution of Occupational Safety and Health (IOSH), for incorporation in systems and procedures.

This stage of the programme takes a number of steps:

- identifying and specifying safe procedures and systems of work;
- explaining to employees, perhaps on a one-to-one basis, the key aspects of the safe system of work and getting them to complete the task according to the established system;
- training new employees in the standard operating procedure;
- supervisors or trained observers use the standard operating procedures to identify employees who are deviating from, or not following correct methods, drawing their attention to same and re-instructing them in the correct methods; and
- regular monitoring of procedures by supervisors, together with employee feedback sessions to discuss improvements in the safe system of work or problems arising from the newly introduced procedure.

Behavioural safety training

Before the start of a behavioural safety programme, it is essential that people at all levels receive the appropriate training. Such a programme incorporates a number of elements, including raising safety awareness, coaching of individuals and reviewing the consequences of unsafe behaviour.

Regular feedback sessions are necessary whereby accidents arising from unsafe behaviour are reviewed with reference to the direct and indirect causes of those accidents. There is no doubt that people learn from other people's mistakes.

Behavioural safety programmes

A well-structured programme incorporates a series of steps:

- Devising a safety observation programme.
- Enlisting and training the observers.
- Devising checklists and, subsequently, observing behaviours.
- Creating a recording system.
- Analysing the data.
- Promoting the programme.
- Implementing the programme.
- Maintaining the behavioural safety process.

Successful behavioural safety programmes

Changing the behaviour of people, in terms of attitude and motivation towards safer working, is no easy task and does not take place overnight. There are many factors which need careful consideration prior to their introduction:

- *Employee participation*
  This is a crucial feature of behavioural safety programmes giving them a say in terms of what they consider to be safe and unsafe behaviour with respect to the processes and activities undertaken at work. Without participation, there is no worker 'ownership', resulting in limited or no commitment to the process.
- *Targeting unsafe behaviours*
  In most cases, only a limited number of unsafe behaviours are identified. However, they may be responsible for a significant number of accidents. By targeting these unsafe behaviours, and the workplace factors that create or drive these behaviours, it is possible to install strategies to eliminate same. These unsafe behaviours must be directly observable by the majority of employees.
- *Data and decision making*
  The old adage, 'What gets measured, gets done!' applies in this case. Trained observers monitor the behaviour of co-employees on a regular basis, producing data. In many cases, the very fact of observing a person's behaviour may bring about changes in that behaviour.

  Data is used in the decision-making process based on identified trends in behaviour and in the provision of feedback to those concerned.

- *Improvements*
  Analysis of data enables the establishment of a schedule of actions that combine to create an overall improvement intervention.
- *Feedback*
  Feedback is the key ingredient of any type of initiative directed at bringing about improvement. Feedback may take many forms, such as simple verbal feedback, graphical feedback, particularly through identifying trends, and by incorporating feedback elements in regular staff briefings.
- *Management support and commitment*
  There must be clear and visible demonstration and commitment to the programme by all levels of management. Without this commitment, the scheme is doomed to failure!
- *The benefits*
  The benefits from the behavioural safety approach can be summarised thus:
  - reduced accidents, incidents, cases of occupational ill health and their associated direct and indirect costs;
  - increased skills in positive reinforcement;
  - increased reporting of hazards, shortcomings in protection arrangements, accidents and near misses;
  - improved levels of safety behaviour and employee commitment to same;
  - acceptance of the system by all concerned (based on the level of management commitment to the system);
  - accelerated action in the case of employee recommendations, suggestions and the remedial action necessary; and
  - improvements in safety culture.

HUMAN ERROR

The HSE publication *Reducing Error and Influencing Behaviour* lists a number of factors that can contribute to human error, which can be a significant causative feature of accidents at work. These include the following.

1. Inadequate information

People do not make errors merely because they are careless or inattentive. Often they have understandable (albeit incorrect) reasons for acting in the way they did. One common reason is ignorance of the production processes in which they are involved and of the potential consequences of their actions.

2. Lack of understanding

This often arises as a result of a failure to communicate accurately and fully the stages of a process that an item has been through. As a result, people make presumptions that certain actions have been taken when this is not the case.

3. Inadequate design

Designers of plant, processes or systems of work must always take into account human fallibility, and never presume that those who operate or maintain plant or systems have a full and continuous appreciation of their essential features. Indeed, failure to consider such matters is, itself, an aspect of human error.

Where it cannot be eliminated, error must be made evident or difficult. Compliance with safety precautions must be made easy. Adequate information as to hazards must be provided. Systems should 'fail safe', that is, refuse to produce unsafe modes of operation.

4. Lapses of attention

The individual's intentions and objectives are correct and the proper course of action is selected, but a slip occurs in performing it. This may be due to competing demands for (limited) attention. Paradoxically, highly skilled performers, because they depend upon finely tuned allocation of their attention, to avoid having to think carefully about every minor detail, may be more likely to make a slip.

5. Mistaken actions

This is the classic situation of doing the wrong thing under the impression that it is right. For example, the individual knows what needs to be done, but chooses an inappropriate method to achieve it.

6. Misperceptions

Misperceptions tend to occur when an individual's limited capacity to give attention to competing information under stress produces *tunnel vision*, or when a preconceived diagnosis blocks out sources of inconsistent information. There is a strong tendency to assume that an established pattern holds good so long as most of the indications are to that effect, even if there is an unexpected indication to the contrary.

One potent source of error in such situations is an inability to analyse and reconcile conflicting evidence deriving from an imperfect understanding of the process itself or the meaning conveyed by the instruments. Full analysis of the preventive measures required involves the need for people to understand the process as well as technical and ergonomic considerations concerned with the instrumentation.

> The official report on the accident in 1979 at the Three Mile Island nuclear power station in the United States cited human factors as the main causes. Misleading and badly presented operating procedures, poor control room design, inadequate training and poorly designed display systems all, in one way or another, gave the operators misleading or incomplete information. In the event, the radioactive exposure off the site was very small indeed. However, the official enquiry emphasised how failures in human factors design, inadequate training and procedures and inadequate management organisation led to a series of relatively minor technical faults being magnified into a near disaster with significant economic and possible human consequences.

7. Mistaken priorities

An organisation's objectives, particularly the relative priorities of different goals, may not be clearly conveyed to, or understood by, individuals. A crucial area of potential conflict is between safety and other objectives, such as output or the saving of cost or time. Misperceptions may then be partly intentional as certain events are ignored in the pursuit of competing objectives. When top management's goals are not clear individuals at any level in the organisation may superimpose their own.

8. Wilfulness

Wilfully disregarding safety rules is rarely a primary cause of accidents. Sometimes, however, there is only a fine dividing line between mistaken priorities and wilfulness. Managers need to be alert to the influences that, in combination, persuade staff to take (and condone others taking) short cuts through the safety rules and procedures because, mistakenly, the perceived benefits outweigh the risks, and they have perhaps got away with it in the past.

The elimination of human error

For the potential for human error to be eliminated or substantially reduced all the above factors need consideration in the design and implementation of safe systems of work, processing operations, work routines and activities. Training and supervision routines should take account of these factors and the various features of human reliability.

PLANNED MOTIVATION SCHEMES

Planned motivation is a method by which the attitudes, and thereby the performance, of people can be improved. Planned motivation schemes are described as an 'industrial catalyst', a tool to maximise performance. They are commonly directed at improving performance in areas such as sales and marketing. However, the concept is also frequently applied to improving safety performance within an organisation. One of the problems with any planned motivation scheme, however, is that, in certain cases the scheme might alter behaviour but not necessarily the attitudes held by employees. On this basis, once the scheme is withdrawn and the rewards are no longer available, people may revert to their original attitudes.

Safety incentive schemes are most effective where:

- people are restricted to one area of activity;
- measurement of safety performance is relatively simple;
- there is regular stimulation or rejuvenation;
- continuing support is provided by management;
- the scheme is assisted by appropriate safety propaganda.

Safety incentive schemes

Safety incentive schemes are a form of planned motivation. With any safety incentive scheme, the main objective is to provide motivation for employees to work safely by identifying targets which can be rewarded if achieved, and making the reward meaningful and desirable to the people concerned. Rewards should not take the form of financial bonuses or payments, however, to the people meeting these targets. On the other hand, formal recognition by the organisation of successfully achieved targets is crucial to the success of any safety incentive scheme.

Important considerations

- Safety incentive schemes should be linked with some form of safety monitoring such as safety inspections or safety sampling exercises.
- Measurable and achievable targets should be set.
- They should definitely not be linked with accident rates as this can result in a tendency not to report minor accidents in particular.
- In terms of lost time, how quickly people return to work after an accident is significant.
- They tend to be short-lived and can get out of hand if they are not regularly monitored and the objectives reinforced at regular intervals.
- If they are not effectively managed, safety incentive schemes can have the effect of shifting management responsibility for safety to employees.

ATYPICAL WORKERS

'Atypical workers' are workers who are not in normal daytime employment, together with shift workers, part-time workers and night workers. Approximately 29 per cent of employees in the United Kingdom work some form of shift pattern, and 25 per cent of employees undertake night shifts. Researchers have, over the years, studied the physical and psychological effects of atypical working on these groups of people, in particular factory workers and transport workers, and have reported a number of findings. For instance:

- Between 60 and 80 per cent of all shift workers experience long-standing sleeping problems.
- Shift workers are 5 to 15 times more likely to experience mood disorders as a result of poor-quality sleep.
- Drug and alcohol abuse are much higher among shift workers.
- 80 per cent of all shift workers complain of chronic fatigue.
- Approximately 75 per cent of shift workers feel isolated from family and friends.
- Digestive disorders are four to five times more likely to occur in shift workers.
- From a safety viewpoint, more serious errors and accidents, resulting from human error, occur during shift work operations.

The psychological factors which affect an individual's ability to make the adjustments required to meet varying work schedules are associated with age, individual sleep needs, sex, the type of work and the extent of desynchronisation of body rhythms, this last point being the most significant.

Reducing the stress of shift work

Strategies are available aimed at minimising the desynchronisation of body rhythms and other health-related effects. The principal objective is to stabilise body rhythms and to provide consistent time cues to the body.

Employers should recognise that workers must be trained to appreciate the stressful effects of shift working, and that there is no perfect solution to this problem. However, they do have some control over how they adjust their lives to the working arrangements and the change in lifestyle that this implies. They need to plan their sleeping, family and social contact schedules in such a way that the stress of this adjustment is minimised. Most health effects arise as a result of changing daily schedules at a rate quicker than that at which the body can adjust. This can result in desynchronisation, with reduced efficiency generally caused by sleep deprivation.

Sleep deprivation

This can have long-term effects on the health of the shift worker. The actual environment in which sleep takes place is important.

Diet

A sensible dietary regime, taking account of the difference between the time of eating and the timing of the digestive system, will assist the worker to minimise discomfort and digestive disorders.

Alcohol and drugs

Avoidance of alcohol and drugs such as caffeine and nicotine can result in improved sleep quality. Occasional use of sleeping tablets may be beneficial, but they should be used under medical supervision.

Family and friends

They should appreciate the demands on the shift worker. Better planning of family and social events is necessary to reduce the isolation frequently experienced by shift workers.

Assisting atypical workers

A number of remedies are available to organisations. These include:

- consultation prior to the introduction of shift work or other forms of atypical working;
- recognition by management that this aspect of work can be stressful for certain groups of workers and of the need to assist in their adjustment to this type of work;
- regular health surveillance of atypical workers to identify any health deterioration or change at an early stage;
- training of shift workers to help them recognise the potentially stressful effects and advise them about the changes in lifestyle that may be needed to reduce these effects;
- better communication between management and workers aimed at reducing the feeling of isolation frequently encountered.

ESTABLISHING A SAFETY CULTURE – THE PRINCIPLES INVOLVED

The main principles involved when establishing a safety culture are generally accepted to be:

(a) the acceptance of responsibility at and from the top, exercised through a clear chain of command, seen to be actual and felt through the organisation;

(b) a conviction that high standards are achievable through proper management;
(c) setting and monitoring of relevant objectives/targets, based upon satisfactory internal information systems;
(d) systematic identification and assessment of hazards and the devising and exercise of preventive systems which are subject to audit and review; in such approaches, particular attention is given to the investigation of error;
(e) immediate rectification of deficiencies; and
(f) promotion and reward of enthusiasm and good results.

(The above list is taken from: Rimington, J R (1989) *The Onshore Safety Regime*, HSE Director General's submission to the Piper Alpha Inquiry.)

Developing a safety culture – essential features

These can be summarised through the guidelines outlined in the CBI's *Developing a Safety Culture*, 1991 (see below).

Several features can be identified from the study which are essential to a sound safety culture. A company wishing to improve its performance will need to judge its existing practices against them:

1. Leadership and commitment from the top which is genuine and visible. This is the most important feature.
2. Acceptance that it is a long-term strategy which requires sustained effort and interest.
3. A policy statement of high expectations and conveying a sense of optimism about what is possible supported by adequate codes of practice and safety standards.
4. Health and safety should be treated as other corporate aims, and properly resourced.
5. It must be a line management responsibility.
6. 'Ownership' of health and safety must permeate all levels of the workforce. This requires employee involvement, training and communication.
7. Realistic and achievable targets should be set and performance measured against them.
8. Incidents should be thoroughly investigated.
9. Consistency of behaviour against agreed standards should be achieved by auditing and good safety behaviour should be a condition of employment.
10. Deficiencies revealed by an investigation or audit should be remedied promptly.
11. Management must receive adequate and up-to-date information to be able to assess performance.

COMMUNICATIONS

Much of the answer lies in effective communication on health and safety. Many staff see, for instance, health and safety training as dull, boring, uninteresting or unrelated to their specific tasks. In certain cases they do not understand the reason why certain precautions are enforced by management and safety practitioners. Furthermore, many safety practitioners lack training in communication, seeing the running of health and safety training operations as a chore. In some cases, they may not have the time due to demands of their full-time job.

As with so many areas of performance, there must be communication both vertically and laterally. The board must set the standards which are both meaningful and measurable. They must communicate these standards down throughout the organisation and ensure such feedback as to enable them to measure and compare performance.

Barriers to communication

A number of barriers can arise at the various phases of the process, in particular:

1. Barriers to reception

Reception of communication can be influenced by:

(a) the needs, anxieties and expectations of the receiver / listener;
(b) the attitudes and values of the receiver; and
(c) environmental stimuli, eg noise.

2. Barriers to understanding

Understanding is a complex process and is affected by:

(a) the use of inappropriate language, technical jargon;
(b) the extent to which the listener can concentrate on receiving the data completely, ie variations in listening skills;
(c) prejudgements made by the listener;
(d) the ability of the listener to consider factors which may be disturbing or contrary to his ideas and opinions, ie the degree of open-mindedness that he possesses;
(e) the length of the communication; and
(f) the degree of knowledge possessed by the listener.

3. Barriers to acceptance

Acceptance of a communication is affected by:

(a) the attitudes and values of the listener;
(b) individual prejudices held by listeners;
(c) status clashes between the sender and the receiver; and
(d) interpersonal emotional conflicts.

HEALTH AND SAFETY TRAINING

Section 2 of the HASAWA places a duty on employers 'to provide such information, instruction, training and supervision as is necessary to ensure, so far as is reasonably practicable, the health and safety at work of his employees'. This duty is extended in subordinate legislation, such as the COSHH Regulations and, in particular, the MHSWR. Regulation 11(2) specifies actual situations and circumstances where health and safety training shall be provided by the employer.

Training was defined by the former Department of Employment as 'the systematic development of attitude, knowledge and skill patterns required by the individual to perform adequately a given task or job. It is often integrated with further education.' The term 'systematic' immediately distinguishes this form of development from the traditional approach consisting most often of the trainee 'sitting by Nellie' and acquiring haphazardly what he could through listening and observation. Systematic training, in effect, makes full utilisation of skills available in training all grades of personnel. It has a number of benefits in that it:

(a) attracts recruits;
(b) achieves the target of an experienced operator's skill in a half or one-third of the traditional time;
(c) creates confidence in trainees that they can acquire diverse skills through application and training;
(d) guarantees better safety performance and morale;
(e) results in greater earnings and productivity;
(f) results in ease, basic mental security and commitment at work;
(g) excludes misfits and diminishes unrest; and
(h) facilitates the understanding and acceptance of change.

Systematic training implies:

(a) the presence of a competent and trained instructor and suitable trainees;
(b) defined training objectives;
(c) a content of knowledge broken down into learnable sequential units;

(d) a content of skills analysed into elements;
(e) a clear and orderly programme;
(f) an appropriate place in which to learn;
(g) suitable equipment and visual aids; and
(h) sufficient time to attain a desired standard of knowledge and competence.

Health and safety training, as with other areas of training, should take place in a number of clearly defined stages.

1. Identification of training needs

A training need is said to exist when the optimum solution to an organisation's problem is through some form of training. For training to be effective it must be integrated to some extent with the selection and placement policies of the organisation. Selection procedures must, for instance, ensure that the trainees are capable of learning what is to be taught.

Training needs should be assessed to cover:

(a) induction training for new recruits;
(b) orientation training of existing employees on, for instance, promotion, change of job, their exposure to new or increased risks, appointment as competent persons, the introduction of new plant, equipment and technology, and prior to the introduction of safe systems of work; and
(c) refresher training directed at maintaining competence.

2. Development of training plan and programme

Training programmes must be co-ordinated with the current human resources needs of the organisation. The first step in the development of a training programme is that of defining the training objectives. Such objectives or aims may best be designed by job specification in the case of new training, or by detailed task analysis and job safety analysis in respect of existing jobs.

3. Implementation of the training plan and programme

Decisions must be made as to the extent of both active and passive learning systems to be incorporated in the programme. Examples of active learning systems are group discussion, role play, syndicate exercises, programmed learning and field exercises, eg safety inspections and audits. Active learning methods reinforce what has already been taught on a passive basis. Passive learning systems incorporate lectures and the use of visual material such as films and videos. With a passive learning system the basic objective must be that of imparting knowledge. The principal advantage of passive learning systems is that they provide frameworks and can be used where large numbers of trainees are involved. Passive learning should be incorporated as an initial introduction to a subject in particular, and should include rules and procedures, providing they are relatively simple.

4. Evaluation of the results

There are two questions that need to be asked at this stage: 'Have the training objectives been met?' and 'If they have been met, could they have been met more effectively?'

Operator training in most industries will need an appraisal of the skills necessary to perform a given task satisfactorily, ie efficiently and safely. It is normal, therefore, to incorporate the results of such an appraisal in the basic training objectives.

A further objective of, particularly, health and safety training is to bring about long-term changes in attitude on the part of trainees, which must be linked with job performance. Any decision, therefore, as to whether training objectives have been met cannot be taken immediately the trainee returns to work or after only a short period of time. It may be several months or even years before a valid evaluation can be made after continuous assessment of the trainee.

The answer to the second question can only be achieved through feedback from personnel monitoring the performance of trainees, and from the trainees themselves. This feedback can usefully be employed in setting objectives for further training, in the revision of training content and in the analysis of training needs for all groups within the organisation.

An important feature of the jobs of health and safety specialists, supervisors and line managers is that of preparing and undertaking short training sessions for their staff on general and specific health and safety issues.

The following points need consideration if such activities are to be successful and effective at conveying the appropriate messages to staff:

1. A list of topics to be covered, eg safe systems of work, manual handling procedures, etc should be developed, and a specific programme should be formulated.
2. Sessions should last no longer than 30 minutes.
3. Visual aids – films, videos, slides, flipcharts, etc – should be used extensively.
4. Topics should, as far as possible, be of direct relevance to the group.
5. Participation should be encouraged with an emphasis on identifying possible misunderstandings or concerns that people may have. This is particularly important when introducing a new safety system or operating procedure.
6. Topics should be presented in a relatively simple fashion, using terms that operators can understand. The use of unfamiliar technical, legal or scientific terminology should be avoided, unless an explanation of such terms is incorporated in the session.
7. Consideration must be given to eliminating any boredom, loss of interest or adverse response on the part of the participants. Talks should be given on as friendly a basis as possible and in a relatively informal atmosphere. Many people respond adversely to a formal classroom situation commonly encountered in staff training activities.

SUMMARY

1. Although most UK health and safety legislation places the duty of compliance firmly on the employer or body corporate, this duty can be discharged by the effective actions of its managers.
2. Studies by the HSE's Accident Prevention Advisory Unit indicate that the vast majority of fatal accidents at work could have been prevented by positive management action.
3. Insufficient attention is paid by managers to the human factors element of safety, in particular those areas of influence on people at work – the organisation, the job and personal factors.
4. Managers need to take into account the organisational characteristics which influence safety-related behaviour.
5. Systems of work and the design of jobs should take into account human capabilities and fallibilities.
6. People are different in terms of how they perceive risk, in their attitude to work, and in what motivates them to work safely.
7. Effective communication and training on health and safety issues is vital in order to ensure good levels of health and safety performance.
8. Increasingly, employers need to consider the concept of behavioural safety and the introduction of behavioural safety programmes.

REFERENCES

CBI (1991) *Developing a Safety Culture*, CBI, London

Health and Safety Executive (1981) *Managing Safety: A review of the role of management in occupational health and safety*, HMSO, London

Health and Safety Executive (1985) *Deadly Maintenance: A study of fatal accidents at work*, HMSO, London

Health and Safety Executive (1985) *Monitoring Safety: An outline report on occupational safety and health by the Accident Prevention Advisory Unit of the Health and Safety Executive*, HMSO, London

Health and Safety Executive (1999) *Reducing Error and Influencing Behaviour* (HSG48), HSE Books, Sudbury

Rimington, J R (1989) *The Onshore Safety Regime*, HSE Director General's submission to the Piper Alpha Inquiry

Stranks, J (1995) *Human Factors and Safety*, Pitman, London

Stranks, J (2007) *Human Factors and Behavioural Safety*, Elsevier Science & Technology, London

6

Ergonomics

There are many definitions of the term 'ergonomics'. These include:

(a) the scientific study of the interrelationships between people and their work;
(b) fitting the task to the individual;
(c) the scientific study of work; and
(d) the study of the man–machine interface.

Considerable attention to the principles of ergonomics, ergonomic design, interface design and anthropometry can have significant benefits in reducing stress in the workforce, thereby promoting greater efficiency and reduced manufacturing losses.

THE APPLICATION OF ERGONOMICS IN WORKING SITUATIONS

Fundamentally, ergonomics covers four principal areas: the human system, the working environment, the man–machine interface and the total working system. Each of these areas is briefly described below.

1. The human system

This area is concerned with people, in particular the physical aspects, such as stamina, strength and body dimensions, and the psychological aspects of human behaviour, such as perception, learning and reaction to given situations. These features are significant in a wide range of jobs.

2. The working environment

Stress in the working environment can have serious effects on worker performance. Typical environmental stressors are extremes of temperature, inadequate or badly designed lighting, inadequate ventilation, high levels of humidity, noise, vibration, dust, fumes and radiation. The provision of a safe working environment, in addition to being a legal requirement under the HASAWA, is a prerequisite for sound levels of worker performance.

3. The man–machine interface

Machine manufacturers and designers frequently produce machinery which places the operator under considerable stress, either through the design and location of controls or badly designed displays. Good standards of design of this man–machine interface take into account the design of controls, displays, the effects of automation and communication systems, particularly on very large machines, with a view to reducing operator error.

4. Total working system

Factors such as the potential for fatigue and stress are considered at this stage, along with aspects such as work rate and productivity. Specific health and safety features are also considered, in particular the effects of operator error.

These factors may be summarised as in Table 6.1.

Human characteristics	*Environmental factors*
Bodily dimensions Strength Stamina Learning Mental and physical limitations Perception Reaction	Temperature Humidity Lighting Noise Vibration Dust, fumes, etc Ventilation
Man–machine interface	*Total working system*
Controls Displays Communications Automation	Work rate Posture Fatigue Stress Productivity Accident and ill health Health and safety

Table 6.1 The ergonomic approach to the work situation

PRINCIPLES OF ERGONOMIC DESIGN

The elimination of operator error is one of the principal objectives of ergonomic design. The following aspects are taken into account in the design process.

1. Vision

The operator should be able to set and read with ease controls, specific instruments and displays of instruments. This prevents or reduces the fatigue that is so frequently the cause of faulty perception and accidents.

2. Posture

Abnormal working postures increase the potential for fatigue, accidents and long-term injury. All work processes and systems of work should be designed to permit a comfortable posture which reduces excessive strain, for example in the case of display screen equipment users and lorry drivers. The siting of controls and displays on a wide range of machinery and plant, vehicles and assembly plant is, therefore, crucial to productivity and the prevention of accidents.

3. Layout

In any working area there should be free movement between operating positions, safe access and egress, and unhindered oral and visual communication. Badly planned, congested and over-populated working areas result in operator fatigue, inattention on the part of the operator and increased accident potential in that area.

4. Comfort

Environmental factors have a direct effect on operator comfort, in particular the levels of lighting and ventilation and the percentage relative humidity of the atmosphere (70 per cent is the optimum relative humidity in most cases).

5. Work rate

Ideally, work rates should be set to suit the operator, but need constant reassessment and revision. Movements which are too slow or too fast cause fatigue. This is particularly applicable where operators are engaged in assembly work using a moving track or conveyor.

CONTROLS AND DISPLAYS

Controls regulate a particular machine function to achieve maximum efficiency. Controls can take many forms, eg the hand brake on a vehicle, the stop–start button of a horizontal lathe. Displays, on the other hand, supply information to the operator, eg a display screen, the petrol gauge in a vehicle. Interface design is concerned with the design and location of controls and displays with the principal objective of minimising or eliminating the potential for operator error. This can be particularly significant in the case of aircraft, large machines, power stations and other complicated machinery and plant where operator error could result in disastrous failure involving loss of life. A number of aspects are important in the design of the man–machine interface:

1. *Separation:* Displays should be separated from physical controls with, preferably, no operational relationship between them.

2. *Order of use:* Controls and displays should be arranged in their order of use, eg left to right for start up and right to left for stopping a machine or closing down the plant.

3. *Comfort:* Where it is not possible to separate controls from displays, they should be mixed to produce a system which can be operated with ease.

4. *Function:* In the case of large consoles, eg in power stations, controls can be divided according to function. Operator training is essential in this case and is aimed at eliminating the chance of error.

5. *Priority:* Here the controls most frequently used are sited in key positions. This form of design is appropriate where there is no competition for space.

6. *Operator fatigue:* Design should take into account the risk of operator fatigue and the likelihood of mistakes being made. Controls should, therefore, be conveniently located with a view to reducing, as far as possible, both visual and postural fatigue.

DISPLAY SCREEN EQUIPMENT

The two principal problems associated with display screen equipment (DSE) operation are visual fatigue and postural fatigue. The need to reduce, as far as possible, these forms of fatigue should preferably be considered when establishing the layout and operation of the DSE workstation.

1. Visual fatigue

This can be associated with:

(a) poor definition of the characters against the background field;
(b) unsuitable background lighting;
(c) glare;
(d) poor legibility, due to factors in the DSE such as 'flicker', 'shimmer' and 'jitter', which are directly related to the refresh rate of the system;
(e) poor-quality source material; and
(f) visual defects on the part of the operator.

Items (a) to (d) can be corrected through good DSE and workplace design. Very few people, however, have perfect vision, the ability to see varying with age and the presence or absence of visual deficiencies such as myopia (short-sightedness) and hypermetropia (long-sightedness). Vision screening is recommended, therefore, as a standard feature of a pre-employment health screen for users.

People who wear spectacles designed for a narrow range of reading distances and those with bifocal or other multifocal lenses may experience some difficulty with tasks involving varying distances. They may find, for instance, that they need to adopt uncomfortable working postures in order to read documents or displayed text satisfactorily. Such individuals may need modifications to their prescription lenses in order to be able to do this type of work, and should consult an optician before undertaking DSE work and whenever discomfort or eye strain is experienced afterwards.

In certain cases, for instance where there is little choice in the positioning of the screen, the use of screen filters may be beneficial. Filters may give a sharper, clearer image, and reduce the eye strain associated with glare and reflection from lights and windows. The use of tinted spectacles, unless prescribed by an optician, is not recommended.

2. Postural fatigue

This is associated with many occupations where staff are frequently sitting for long periods. Postural fatigue can be experienced by way of neck, shoulder and back ache, together with persistent headaches. The incorporation of the recommendations that follow should alleviate the problem of postural fatigue.

3. Operational stress

The extent of stress experienced by operators will vary, but the following practices are recommended:

(a) rotation of DSE users in order to take them away from the screen for limited periods; and
(b) trainees should have a short break from DSE operation every one to two hours during their first three months of training.

It should be appreciated that a good logical screen layout can be helpful in the alleviation of operational stress.

4. Work-related upper limb disorders

Work-related upper limb disorders (WRULD) associated with repetitive strain injuries (RSI) have a history going back to the 18th century. RSI is a term covering a range of conditions such as tenosynovitis, carpal tunnel syndrome and tennis elbow, and is associated with the tasks people perform in work situations. These tasks may involve over-forceful and repetitive movements, inadequate rest periods, excessive workloads and sustained or constrained postures. Typical occupations affected by RSI are keyboard operators, supermarket checkout personnel, assembly workers; in effect, anyone involved with continuous and repetitive movements during work, such as bricklayers.

The resulting effects on health may include muscular inflammation and tenosynovitis, an inflammatory condition of the synovial lining of the tendon sheath. Tenosynovitis has been a prescribed occupational disease for a number of years, but the HSE changed the name of the condition to 'work-related upper limb disorder' because the disorder does not always result from strain or repetition and is not always a visible injury.

True tenosynovitis is rare, and has serious consequences in most cases. The more common and benign form is peritendinitis crepitans. This is associated with inflammation of the joint between a tendon and a muscle, often extending well into the muscle tissue. The clinical signs and symptoms can include crepitus (a grating sensation in the joint), tenderness, aching and inflammation, which is aggravated by movement or the application of manual pressure.

RSI can take a number of forms:

- *Epicondylitis.* This is an inflammatory condition of the area of the arm where a muscle joins a bone, resulting in pain and swelling of the joint, as in tennis elbow.
- *Carpal tunnel syndrome.* Nerves and blood vessels pass through the carpal bone at the base of the hand. With this condition there is pain in the hand and wrist because of fluid and tissues pressing on a wrist nerve, resulting eventually in tingling and numbness in the fingers and weakness of the hand.

- *Peritendinitis.* This is an inflammatory condition of the area where a tendon joins a muscle. It results in pain in the wrist and forearm, together with localised swelling.
- *Tendinitis.* This condition involves inflammation of a tendon or tendons, particularly in the fingers. It causes aches, numbness or tenderness. In some cases the movement of arm, hand and fingers is affected.
- *Tenosynovitis.* This involves inflammation of the synovial lining of the tendon sheath.
- *Writer's cramp.* Many people will have experienced this condition briefly at some time in their lives. It results in painful cramps in the hand, forearm and fingers.
- *Dupuytren's contracture.* This condition is associated with thickening of the tissue below the palm of the hand, with the result that it is impossible to straighten the hand and fingers.
- *Trigger finger and trigger thumb.* Here the finger or thumb 'locks' before jerking into position. The condition can be extremely painful.

In most cases there is a need to pursue ergonomic solutions, such as improving workstation layout, which may include job rotation, whereby people are taken away from the particular repetitive task for periods of time.

Health and Safety (Display Screen Equipment) Regulations 1992

Poor design and layout of the workstation and lack of attention to environmental factors, eg background lighting, are the principal causes of occupational stress, ie visual and postural fatigue, discomfort and, for some people, annoyance. In certain cases, this lack of attention can lead to one or more work-related upper limb disorders.

The Health and Safety (Display Screen Equipment) Regulations 1992 apply to users and operators of this equipment. A 'user' is an employee who *habitually* uses display screen equipment as a significant part of his *normal* work. An 'operator', on the other hand, is a self-employed person who habitually uses display screen equipment as a significant part of his *normal* work. Typical examples of display screen equipment users are word-processing pool workers, secretaries, data input operators, journalists, tele-sales operators and graphic designers. A receptionist in a hotel would not be classified as a 'user'.

The regulations, which are accompanied by an HSE Guidance Note, place specific duties on employers. Thus, each employer shall:

(a) perform a suitable and sufficient analysis of workstations used by users and operators; (Regulation 2)
(b) ensure that new workstations meet the requirements laid down in the schedule to the regulations; (Regulation 3)

Figure 6.1 DSE workstation principles
(Source: Health and Safety (Display Screen Equipment) Regulations 1992)

(c) plan the activities of users so that their daily work on display screen equipment is periodically interrupted by breaks or changes of work activity to reduce their workload at that equipment; (Regulation 4)
(d) provide eye and eyesight tests for users and persons who are to become users, which must be carried out by a competent person; (Regulation 5)
(e) provide adequate health and safety training for users in the use of any workstation upon which he may be required to work; (Regulation 6) and
(f) provide adequate information to ensure that operators and users know about all aspects of health and safety relating to their workstations, and about such measures taken by the employer in compliance with his duties under Regulations 2 and 3 as relate to them and their work.

A 'workstation' is defined as an assembly comprising:

(a) display screen equipment (whether provided with software determining the interface between the equipment and its operator or user, a keyboard or any other input device);
(b) any optional accessories to the display screen equipment;

(c) any disk drive, telephone, modem, printer, document holder, work chair, work desk, work surface or other item peripheral to the display screen equipment; and
(d) the immediate environment around the display screen equipment.

The analysis of workstations should be based on the criteria outlined in the schedule below.

THE SCHEDULE

This schedule sets out the minimum requirements for workstations, which are contained in the annex to Council Directive 90/270/EEC on the minimum safety and health requirements for work with display screen equipment.

1. Extent to which employers must ensure that workstations meet the requirements laid down in this schedule

An employer shall ensure that a workstation meets the requirements laid down in this schedule to the extent that:

(a) those requirements relate to a component which is present in the workstation concerned;
(b) those requirements have effect with a view to securing the health, safety and welfare of persons at work; and
(c) the inherent characteristics of a given task do not make compliance with those requirements inappropriate as respects the workstation concerned.

2. Equipment

General comment
The use as such of the equipment must not be a source of risk for operators or users.

Display screen
The characters on the screen shall be well defined and clearly formed, of adequate size and with adequate spacing between the characters and lines.

The image on the screen should be stable, with no flickering or other forms of instability.

The brightness and contrast between the characters and the background shall be easily adjustable by the user, and also easily adjustable to ambient conditions.

The screen must swivel and tilt easily and freely to suit the needs of the operator or user.

It shall be possible to use a separate base for the screen or an adjustable table.

The screen shall be free of reflective glare and reflections liable to cause discomfort to the user.

Keyboard

The keyboard shall be tiltable and separate from the screen, so as to allow the operator or user to find a comfortable working position, avoiding fatigue in the arms or hands.

The space in front of the keyboard shall be sufficient to provide support for the hands and arms of the operator or user.

The keyboard shall have a matt surface to avoid reflective glare.

The arrangement of the keyboard and the characteristics of the keys shall be designed to facilitate the use of the keyboard.

The symbols on the keys shall be adequately contrasted and legible from the design working position.

Work desk or work surface

The work desk or work surface shall have a sufficiently large, low-reflectance surface and allow a flexible arrangement of the screen, keyboard, documents and related equipment.

The document holder shall be stable and adjustable and shall be positioned so as to minimise the need for uncomfortable head and eye movements.

There shall be adequate space for operators or users to find a comfortable position.

Work chair

The work chair shall be stable and allow the user easy freedom of movement and a comfortable position.

The seat shall be adjustable in height.

The seat back shall be adjustable in both height and tilt.

A footrest shall be made available to any user who wishes one.

3. Environment

Space requirements

The workstation shall be designed to provide sufficient space for the user to change position and vary movements.

Lighting

Any room lighting or task lighting provided shall ensure satisfactory conditions and an appropriate contrast between the screen and the background environment, taking into account the type of work and the vision requirements of the operator or user.

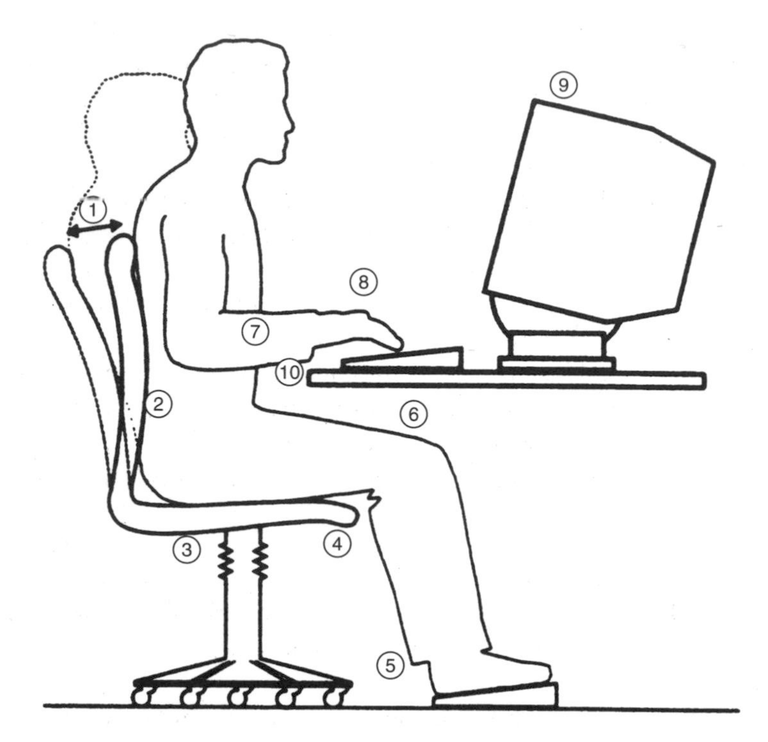

Figure 6.2 Correct seating and posture
(Source: Health and Safety (Display Screen Equipment) Regulations 1992)

Possible disturbing glare and reflections on the screen or other equipment shall be prevented by co-ordinating workplace and workstation layout with the positioning and technical characteristics of the artificial light sources.

Reflections and glare

Workstations shall be so designed that sources of light, such as windows and other openings, transparent or translucid walls, and brightly coloured fixtures on walls cause no direct glare or distracting reflections on the screen.

Windows shall be fitted with a suitable system of adjustable covering to attenuate the daylight that falls on the workstation.

Noise

Noise emitted by equipment belonging to any workstation shall be taken into account when a workstation is being equipped, with a view in particular to minimising distraction.

Heat

Equipment belonging to any workstation shall not produce excess heat which could cause discomfort to operators or users.

Radiation
All radiation with the exception of the visible part of the electromagnetic spectrum shall be reduced to negligible levels from the point of view of the protection of operators' or users' health and safety.

Humidity
An adequate level of humidity shall be established and maintained.

4. Interface between computer and operator/user

In designing, selecting, commissioning and modifying software, and in designing tasks using display screen equipment, the employer shall take into account the following principles:

(a) software must be suitable for the task;
(b) software must be easy to use and, where appropriate, adaptable to the user's level of knowledge or experience; no quantitative or qualitative checking facility may be used without the knowledge of the operators or users;
(c) systems must provide feedback to operators or users on the performance of those systems;
(d) systems must display information in a format and at a pace adapted to operators or users;
(e) the principles of software ergonomics must be applied, in particular to human data processing.

MANUAL HANDLING INJURIES AND CONDITIONS

People involved in manual handling operations can sustain both external and internal injuries. External injuries are common and can include bruising and lacerations to hands, fingers, forearms, ankles and feet, cuts to hands and finger-crushing injuries.

The more serious internal injuries include hernias (ruptures), prolapsed intervertebral discs, and ligamental and muscle strains. It is these internal injuries that result in manual handling injuries being the principal cause of lost time from work. There is a need, therefore, to ensure these injuries and conditions do not arise, through the provision of instruction and training in safe manual handling techniques.

Hernia (rupture)

A person suffers a rupture when a body organ or part of an organ, such as a loop of the small intestine, protrudes from one compartment of the body into another. In the case of the small intestine, this protrusion may be from the abdominal cavity into the groin, or through the frontal abdominal wall. In both cases this may result from incorrect or thoughtless handling techniques, and in particular, keeping the back bent when lifting.

The adoption of a bent back stance results in compression of the abdomen and small intestine, putting considerable pressure on the small gap in the abdominal muscles where (in men) the testis descends to the scrotum. Excessive straining, and even hard coughing, may cause a bulge at the gap, and a loop of intestine or other abdominal structure can easily slip through the gap. This is known as an inguinal hernia.

Most inguinal hernias are reasonably well coped with, but the hernia may become strangulated, whereby the loop of intestine is nipped or pinched at the entrance to the hernia. This results in obstruction of the intestine, and fresh blood no longer reaches this point. Swift medical attention is needed at this stage.

Slipped discs

The human spine comprises a number of vertebrae which are, fundamentally, small interlocking bones that permit bending of the back. The vertebrae are separated by a disc composed of both fibrous and cartilaginous material, an intervertebral disc, which has a shock-absorbing function to protect the complete spine.

A prolapsed or slipped disc occurs when an intervertebral disc becomes displaced from its original position and no longer performs its function. In some cases there may be compression of a disc. This results in an extremely painful condition, commonly associated with sciatica in the left or right leg, and sometimes leading to partial paralysis.

The most common cause of a slipped disc is through incorrect lifting, that is, bending the back while lifting, as opposed to keeping the back reasonably straight and using the strong muscles in the legs to raise a load.

Ligamental and muscular strain

Ligaments are fibrous bands occurring where two bones form a joint, such as at the knee and elbow. While ligaments are flexible to some extent, they do not stretch. They come into play at the extremes of movement, and set the limits beyond which no further movement is possible in a joint. Where a joint is forced beyond its limit, the ligament tears, resulting in a sprain.

Torn ligaments can arise from jerky handling movements that put stress on a joint, in badly co-ordinated team lifting, and when someone lets go of a load halfway through a lift after realising it is too heavy.

When muscles are brought into use for manual handling operations, they are subjected to varying degrees of stress. In most cases, carrying a load imposes a strain on many groups of muscles, particularly those in the trunk and arms. However, the muscles of the back are designed for static work only and not for dynamic work of this kind, and where bent back stances are used, fatigue very soon arises, accompanied by pain in the back muscles. Hence the need to use the arm and leg muscles, which perform dynamic work, when undertaking manual handling activities.

Rheumatism

A secondary outcome of manual handling operations for many people is one of the various forms of rheumatism, such as rheumatoid arthritis, lumbago, fibrositis, osteoarthritis and rheumatic fever. These conditions are associated with painful disorders of joints and muscles, particularly as people get older.

Manual handling instruction should draw the attention of trainees to the fact that the damage they do to muscles, ligaments and joints in their early working life through careless handling of loads may well manifest itself in various forms of rheumatism in later life.

Principles of correct handling

1. Correct grip

The correct grip makes use of the palm of the hand and roots of the fingers and thumb. (See Figure 6.3.) Do not grip with the fingertips as this will lead to strained fingers and muscles in the forearm.

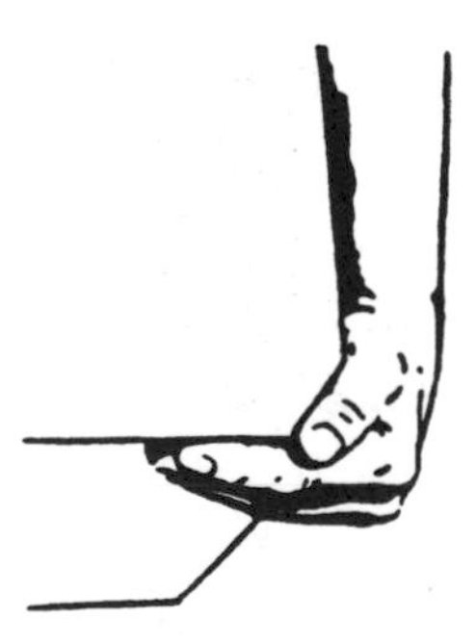

Figure 6.3 Correct handling – proper grip

2. Straight back

In order to pick up a load with a straight back one must approach the task by flexing the hips, knees and ankles and the load must be held close to the body.

The lift is brought about by the powerful muscles of the legs and not the back, which is kept straight throughout the movement. (See Figure 6.4.)

Remember! A bent back is a weak back and can lead to a strained back.

Figure 6.4 Correct handling – straight back

3. Head up

Practise raising the top of the head and this will help you maintain a straight back – an essential movement that should be carried out prior to every lift. This will also enable you to see where you are going.

4. Correct foot position

Always have the feet apart but no wider than the hips, and one foot should be in advance of the other. This leading foot should point in the direction you intend to move. (See Figure 6.5.)

Figure 6.5 Correct handling – proper foot positions

5. Arms close to the body

Lifting, carrying or pushing with the arms away from the sides of the body result in needless strain being put on the chest, upper back and shoulder muscles. Keep the arms as close to the body as possible.

6. Use your body weight

Properly employed, your body weight can be used in moving a load by acting as a counterbalance and thus reducing the amount of muscular effort. (See Figure 6.6.)

Note: These basic principles should be applied in all manual handling operations. They do, however, need regular practice and training of operators in the techniques.

Figure 6.6 Correct handling – use your bodyweight

Manual Handling Operations Regulations 1992

These regulations implement the European 'Heavy Loads' Directive. They supplement the general duties placed upon employers and others by the HASAWA and the broad requirements of the MHSWR 1992, and are supported by Guidance issued by the HSE.

The regulations define a number of terms as follows:

Injury – does not include injury caused by any toxic or corrosive substance which:

(a) has leaked or spilt from a load;
(b) is present on the surface of a load but has not leaked or spilt from it; or
(c) is a constituent part of a load.

Load – includes any person and any animal.

Manual handling operations – means any transporting or supporting of a load (including the lifting, putting down, pushing, pulling, carrying or moving thereof) by hand or by bodily force.

The following duties of employers are laid down under Regulation 4:

1. Each employer shall:

(a) so far as is reasonably practicable, avoid the need for his employees to undertake any manual handling operations at work which involve a risk of their being injured;
(b) where it is not reasonably practicable to avoid the need for his employees to undertake any manual handling operations at work which involve a risk of their being injured, he must:
    (i) make a suitable and sufficient assessment of all such manual handling operations to be undertaken by them, having regard to the factors which are specified in the schedule;
    (ii) take appropriate steps to reduce the risk of injury to those employees arising out of their undertaking any such manual handling operations to the lowest level reasonably practicable;
    (iii) take appropriate steps to provide any of those employees who are undertaking such manual handling operations with general indications and, where it is reasonably practicable to do so, precise information on:
        (aa) the weight of each load, and
        (bb) the heaviest side of any load whose centre of gravity is not positioned centrally.

2. Any assessment such as referred to in para 1(b)(i) of this regulation shall be reviewed by the employer who made it if:

(a) there is reason to suspect it is no longer valid; or
(b) there has been a significant change in the manual handling operations to which it relates.

If changes to an assessment are required, as a result of any such review, the relevant employer shall make them.

The regulations further require that each employee, while at work, shall make full and proper use of any system of work provided for his use by his employer in compliance with Regulation 4(1)(b)(ii) of these Regulations.

3. In determining for the purpose of this regulation whether manual handling operations involve a risk of injury and in determining the appropriate steps to reduce that risk regard shall be had in particular to:

(a) the physical suitability of the employee to carry out the operations;

(b) the clothing, footwear and other personal effects he is wearing;
(c) his knowledge and training;
(d) the results of any relevant risk assessment carried out pursuant to Regulation 3 of the MHSWR;
(e) whether the employee is in a group of employees identified by that assessment as being especially at risk; and
(f) the results of any health surveillance provided pursuant to Regulation 6 of the MHSWR.

Schedule 1

Factors to which the employer must have regard and questions he must consider when making an assessment of manual handling operations:

1. The tasks
Do they involve:

- holding or manipulating loads at distance from trunk?
- unsatisfactory bodily movement or posture, especially:
  - twisting the trunk?
  - stooping?
  - reaching upwards?
- excessive movement of loads, especially:
  - excessive lifting or lowering distances?
  - excessive carrying distances?
- excessive pushing or pulling of loads?
- risk of sudden movement of loads?
- frequent or prolonged physical effort?
- insufficient rest or recovery periods?
- a rate of work imposed by a process?

2. The loads
Are they:

- heavy?
- bulky or unwieldy?
- difficult to grasp?
- unstable, or with contents likely to shift?
- sharp, hot or otherwise potentially damaging?

3. The working environment
Are there:

- space constraints preventing good posture?
- uneven, slippery or unstable floors?
- variations in level of floors or work surfaces?

- extremes of temperature or humidity?
- conditions causing ventilation problems or gusts of wind?
- poor lighting conditions?

4. Individual capability

Does the job:

- require unusual strength, height, etc?
- create a hazard to those who might reasonably be considered to be pregnant or have a health problem?
- require special information or training for its safe performance?

5. Other factors

Is movement or posture hindered by personal protective equipment or clothing?

MANAGEMENT ACTION

Management should consider an action plan aimed at eliminating or reducing accidents and back pain associated with poor manual handling practices. Such an action plan should incorporate the following:

1. Identification of all manual handling operations and the risks to operators.
2. Examination of current sickness absence records associated with manual handling, eg days lost through back strain, fatigue, handling injuries, and identification of the costs to the organisation.
3. Examination of the nature of the loads handled by people in terms of size, shape, weight, rigidity, etc.
4. Consideration of the availability of existing manual handling equipment and the need for improved equipment in the future.
5. Take into account the physical and mental characteristics of future staff in terms of their ability to handle loads as a feature of pre-employment health screening.
6. Undertake training of staff, both at the induction stage and on a regular basis, and highlight the risk of back injury through the use of posters, films and other forms of increasing awareness.

SUMMARY

1. Ergonomics is concerned with people and the work that they do, the potential for error in their work and the effects of work on their health.
2. Good standards of ergonomic design are essential in order to reduce operator error.

3. Design of the 'man–machine interface' should consider, particularly, the use and layout of controls and displays.
4. The introduction of the 'new technology' in offices has created a number of problems in terms of operator visual fatigue and postural fatigue. DSE workstation design should consider these aspects with a view to their prevention.
5. Over 25 million working days are lost through back pain. Back pain is the second most common reason, next to the common cold, for people being absent from work. The need for well-established manual handling procedures is, therefore, paramount if this loss to the country is to be reduced.

REFERENCES

Bell, C R (1974) *Men at Work*, George Allen & Unwin Ltd, London

Health and Safety Executive (1992) *Health and Safety (Display Screen Equipment) Regulations 1992 and Guidance on Regulations*, HMSO, London

Health and Safety Executive (1992) *Lighten the Load: Guidance for employees on musculoskeletal disorders*, HSE Information Centre, Sheffield

Health and Safety Executive (1992) *Manual Handling Operations Regulations 1992 and Guidance on Regulations*, HMSO, London

Health and Safety Executive (1993) *Ergonomics at Work*, HSE Information Centre, Sheffield

Health and Safety Executive (1993) *Working with VDUs*, HSE Information Centre, Sheffield

Health and Safety Executive (1995) *Upper Limb Disorders: Assessing the risks*, HSE Books, Sudbury

Health and Safety Executive (1999) *Checkouts and Musculoskeletal Disorders*, HSE Books, Sudbury

Health and Safety Executive (2000) *Getting to Grips with Manual Handling*, HSE Information Centre, Sheffield

Health and Safety Executive (2001) *Power Tools: How to reduce vibration health risks, Guide for employers* (INDG338), HSE, London

International Labour Organisation (1984) *ILO Encyclopaedia: Lifting and carrying*, ILO, Geneva

McCormick, E J (1976) *Human Factors Engineering*, McGraw-Hill, New York

Shackel, B (1974) *Applied Ergonomics Handbook*, IPC Science and Technology Press, Guildford

Singleton, W T (1974) *Man-Machine Systems*, Penguin Books, Harmondsworth

Singleton, W T (1976) *Human Aspects of Safety*, Keith Shipton Developments, London

7

Stress at Work

The amount of time lost from work in the United Kingdom is typically in excess of 300 million days per annum. At least 10 per cent of these days lost are attributed to what is officially referred to as 'psychoneurosis'. To this figure of 30 million working days lost through psychoneurosis must be added further time lost through 'psychosomatic' complaints, namely physical illnesses that originated in, or have been exacerbated by, psychological and stress-related problems, and all uncertified absence. Such losses represent a significant cost to all forms of organisation, and do not include the cost of low productivity and decreased efficiency due to low motivation, increases in alcohol and drug consumption, and time lost through 'presenteeism' ie being physically present at work, but mentally absent.

WHAT IS STRESS?

Stress is a term that is rarely clearly understood. Various definitions have been put forward over the years, as follows:

1. Any influence that disturbs the natural equilibrium of the living body.
2. The common response to attack (Hans Selye, 1936).
3. A feeling of sustained anxiety which, over a period of time, leads to disease.
4. A psychological response which follows failure to cope with problems.

Generally, a stressful circumstance is one with which an individual is unable to cope successfully, or believes he cannot cope successfully, and which results in unwanted physical, mental or emotional responses. Stress implies some form of demand on the individual, it can be perceived as a threat, it can produce the classic 'flight or fight' response, it may create physiological imbalance and can certainly affect individual performance. It is particularly concerned with how people cope with changes in their lives at work, at home and in other circumstances. It should be appreciated, however, that not all stress is bad. We all need a certain amount of stress (positive stress) in order to cope with life situations on an ongoing basis.

Classification of stressors

A stressor produces stress. There are many forms of stressor, namely:

1. Physical stressors – extremes of temperature, lighting, ventilation and humidity, noise and vibration.
2. Chemical stressors – dangerous chemicals: gases, vapours, dusts, etc.
3. Biological stressors – bacteria, viruses, etc.

However, most people associate stress with social or psychological stress which may be brought about, perhaps, by isolation, rejection, pressure and a general overloading of the body systems (distress). The demands on people at work vary substantially. Some demands may be related to the actual work that they do or the factors surrounding that work, including:

(a) psychological demands – machine-paced work, the quality of supervision, hazards, monotony of the task;
(b) physical demands – the effort required, as in manual handling activities, the potential for fatigue and exposure to hazardous substances;
(c) demands related to the construction of displays and controls on machinery – display screen equipment, fork-lift trucks, machinery;
(d) environmental demands – noise, pollution, poor lighting, etc;
(e) working hours – shift work, unsocial hours, night work, the frequency of breaks; and
(f) payment arrangements – piecework systems, compliance with quality standards.

Sources of stress among managers may be associated with many factors – their role in the organisation, career development, the organisational structure and climate, relationships within the organisation and certain factors which are specific to the job. There may also be demands from outside the organisation. Cooper and Marshall (1978) demonstrated the conflict between the demands of the organisation and external demands, eg those of the family, as a significant cause of management stress (see Figure 7.1). The concept of the

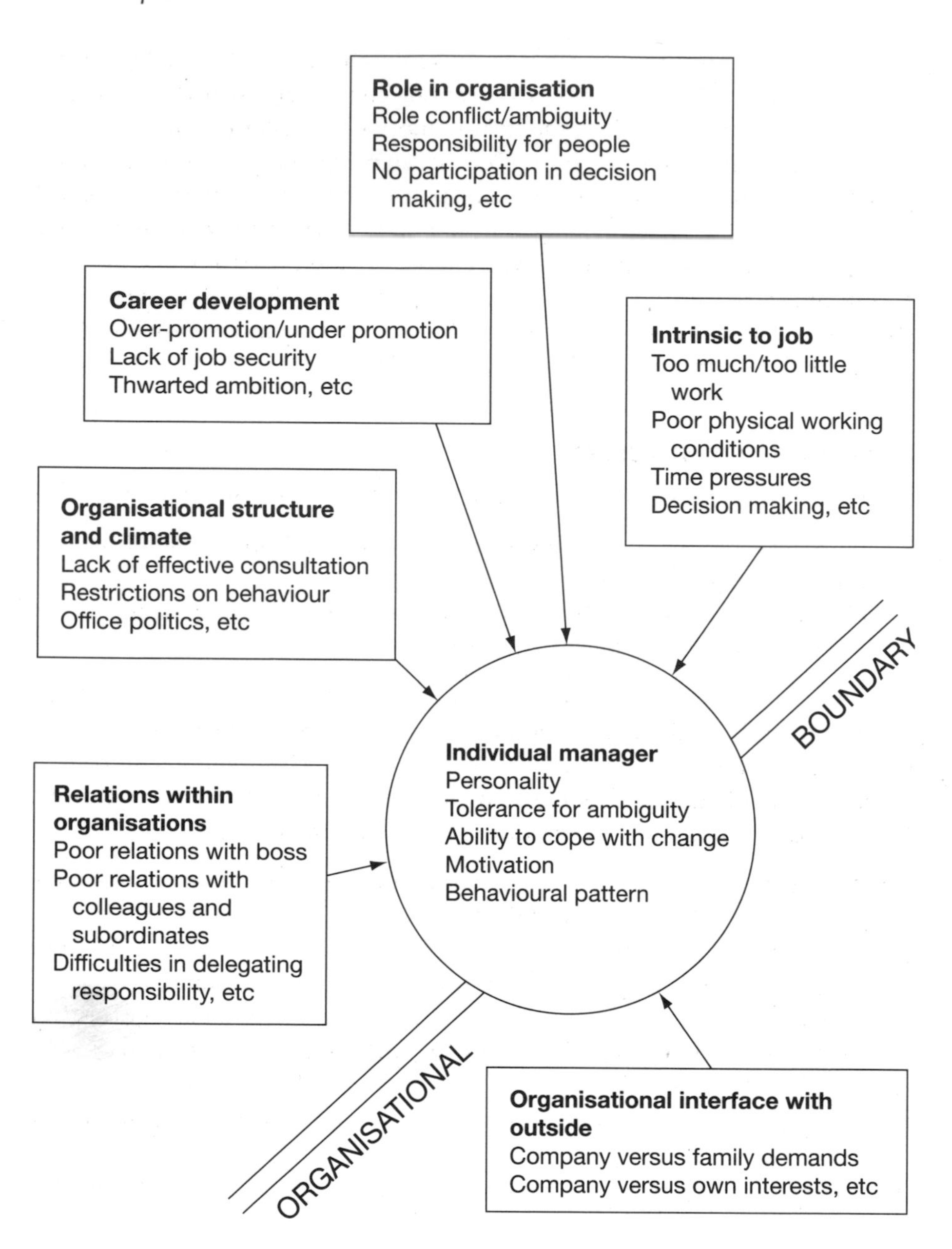

Figure 7.1 Sources of managerial stress

(Source: Cooper and Marshall, 1978)

organisational boundary, with the manager astride that boundary, is well established. Personal factors are important in this case in terms of individual personality, a person's tolerance for ambiguity, ability to cope with change, level of motivation and specific behavioural patterns.

Role theory

Role theory views most large organisations as comprising systems of interlocking roles. These roles relate to what people do and what others expect of them. Problems arise as a result of:

1. **Role ambiguity** This is the situation where the role holder has insufficient information for the adequate performance of his role, or where the information received is open to more than one interpretation. Potentially ambiguous situations are in posts where there is a time lag between action taken and visible results, or where the role holder is unable to see the results of his actions.
2. **Role conflict** This arises where members of the organisation, who exchange information with the role holder, have different expectations of his role. (Health and safety specialists frequently suffer this conflict situation.) Each may exert pressure on the role holder and, commonly, satisfying one expectation could make compliance with the other expectations difficult. This is the classic 'servant of two masters' situation.
3. **Role overload** This results from a combination of role ambiguity and role conflict. The role holder works harder to clarify normal expectations or to satisfy conflicting priorities which are frequently impossible to achieve within the time limits specified.

Research has shown that when experience of role conflict, ambiguity and overload is high, then job satisfaction is low. This may well be coupled with worry and anxiety. These factors may add to the onset of stress-related diseases and conditions such as peptic ulcers, coronary heart disease and nervous breakdowns.

Personality and stress

No two people necessarily respond to the same stressor in the same way. Individual personality factors are significant. Personality was defined by Allport as: 'the dynamic organisation within the individual of the psychophysical systems that determine his characteristic behaviour and thought'. Various types and traits of personality have been established over the last 30 years; these are classified as follows:

1. Type 'A' – Ambitious	Active and energetic; impatient if they have to wait in a queue; conscientious; maintain high standards; time is a problem – there is never enough; frequently intolerant of those who may be slower in thought or action.
2. Type 'B' – Placid	Quiet; very little worries them; put their worries into things they can alter or control and leave others to worry about the rest.
3. Type 'C' – Worrying	Nervous; highly strung; not very confident of self-ability; anxious about the future and of being able to cope.
4. Type 'D' – Carefree	Loves variety; often athletic and daring; very little worries them; not concerned about the future.
5. Type 'E' – Suspicious	Dedicated and serious; very concerned with other people's opinions of them; do not take criticism kindly and tend to dwell on such criticism for a long time; distrust most people.
6. Type 'F' – Dependent	Bored with their own company; sensitive to surroundings; rely on others a great deal; people who interest them are oddly unreliable; they find that the people they really need are boring; do not respond easily to change.
7. Type 'G' – Fussy	Punctilious; conscientious and like a set routine; do not like change; any new problem throws them because there are no rules to follow; conventional and predictable; great believers in authority.

Research indicates that most people combine traits of more than one of these 'types', and so the above definitions can only be used as a guide. The type most at risk to stress is Type A.

Women at work

Women can be subject to many stressors at work which are not suffered by their male counterparts. While sexual harassment is a common cause of stress among women, other causes of stress include:

(a) performance-related pressures;
(b) lower rates of pay;
(c) the problem of maintaining dependants at home;
(d) lack of encouragement from superiors, including not being taken seriously;

(e) discrimination in terms of advancement;
(f) sex discrimination and prejudice;
(g) pressure from dependants at home;
(h) career-related dilemmas, including whether to start a family or whether to marry or live with someone;
(i) lack of social support from colleagues;
(j) being single and labelled as an oddity; and
(k) lack of domestic support at home.

Management should be aware of the various forms of stress women are exposed to while at work. Wherever necessary measures should be taken to reduce stress. Special attention should be given to cases of sexual harassment, which can be handled using disciplinary action.

The effects of stress

Stress effects vary considerably from person to person. Typical effects of stress are headaches, insomnia, fatigue, overeating, constipation, nervousness, minor accidents, palpitations, indigestion and irritability. Many more effects and symptoms could be added to this list.

The two principal psychological effects of stress are anxiety and depression.

1. **Anxiety** This is a state of tension coupled with apprehension, worry, guilt, insecurity and a constant need for reassurance. It is accompanied by psychosomatic symptoms, such as profuse perspiration, difficulty in breathing, gastric disturbances, rapid heart beat, frequent urination, muscle tension or high blood pressure. Insomnia is a reliable indicator of a state of anxiety.

2. **Depression** This has been defined as 'a sadness which has lost its relationship to the logical progression of events' (American psychiatrist David Viscott). Its milder form may be a direct result of a crisis in work relationships. Severe forms may exhibit biochemical disturbances, and the extreme form can lead to suicide. Another definition is 'a mood, characterised by feelings of dejection and gloom, and other permutations, such as feelings of hopelessness, futility and guilt'.

Typical stressful conditions

1. Too heavy or too light a workload.
2. A job which is too difficult or too easy.
3. Working excessive hours, eg 60 or more hours per week.
4. Conflicting job demands – the 'servant of two masters' situation.
5. Too much or too little responsibility.
6. Poor human relationships.

7. Incompetent superiors, in terms of their ability to make decisions, their level of performance and their job knowledge.
8. Lack of participation in decision making and other activities where a joint approach would be beneficial.
9. Middle-age vulnerability associated with reduced career prospects or the need to change career, the threat of redundancy or premature retirement.
10. Over-promotion or under-promotion.
11. Interaction between work and family commitments.
12. Deficiencies in interpersonal skills.

Coping strategies

There are a number of ways that people can commonly deal with the emotional and physical aspects of stress at work. These are generally through relaxation training, physical exercise and, in certain cases, the use of drugs. Relaxation training may take the form of progressive muscular relaxation, brief relaxation exercises and several forms of meditation, eg mantra meditation.

From an organisational viewpoint, stress at all levels can have a serious effect on performance. Potentially stressful organisations are those:

(a) which are large and bureaucratic;
(b) in which there are formally prescribed rules and regulations;
(c) where there is conflict between positions and people;
(d) where people are expected to work hard for long hours;
(e) where no praise is given;
(f) where the general culture is classed as 'unfriendly'; and
(g) where conflict can arise between normal work and external interests, eg the family.

There is a need, therefore, for organisations to do certain things if they are to reduce stress in the workplace. A stress management action plan should incorporate the following:

1. Recognition of the causes and symptoms of stress at all levels.
2. Decisions on the need to do something about it.
3. Identification of the group or groups who may be affected by stress at work.
4. Examination and evaluation by interview or questionnaire to determine the causes of stress.
5. Analysis of problem areas.
6. Decision on appropriate strategies, eg training, time management, counselling of and support for individuals, revision of management policies in certain cases.
7. Implementation of a stress management programme taking the above factors into account.

HSE management standards

The HSC has introduced a programme of work to tackle occupational stress through a range of actions, including the development of good standards of management practice. Based on the responses to a former discussion document and the results of a research programme, the HSC concluded that:

- work-related stress is a serious problem;
- work-related stress is a health and safety issue;
- it can be tackled in part through the application of health and safety legislation.

However, in the absence of any clear standards of management practice against which an employer's performance in managing a range of stressors, such as the way the work is structured, could be measured, the HSC asked the HSE to develop standards of management practice for controlling work-related stressors.

The management standards are aimed at those stressors that affect the majority of employees in an organisation, and cover six main factors which can lead to work-related stress. The standards are outlined below.

1. **Demands.** At least 85 per cent of employees indicate that they are able to cope with the demands of their jobs, and systems are in place locally to respond to any individual concerns.
2. **Control.** At least 85 per cent of employees indicate they are able to have a say about the way they do their work, and systems are in place locally to respond to any individual concerns.
3. **Support.** At least 85 per cent of employees indicate that they receive adequate information and support from their colleagues and superiors, and systems are in place locally to respond to any individual concerns.
4. **Relationships.** At least 65 per cent of employees indicate that they are not subjected to unacceptable behaviours (eg bullying) at work, and systems are in place locally to respond to any individual concerns.
5. **Role.** At least 65 per cent of employees indicate that they understand their role and responsibilities, and systems are in place locally to respond to any individual concerns.
6. **Change.** At least 65 per cent of employees indicate that the organisation engages them frequently when undergoing an organisational change, and systems are in place locally to respond to any individual concerns.

Each of these standards is accompanied by a number of criteria indicating the state to be achieved to reach that standard.

VIOLENCE AT WORK

What is violence?

Most people would associate violence with being physically attacked by another person or group of persons resulting in, in extreme cases, death and, in other cases, physical injury. Physical injury may take the form of cuts, abrasions, bruising, fractures, dislocation of joints, total or partial blindness; all of which may require either first aid treatment or treatment at the accident and emergency department of a local hospital. Physical disability may further arise as a result of violence at work.

However, violence may not necessarily result in physical injury alone. It may take the form of psychological violence arising from verbal threats, persistent verbal abuse, bullying, obstruction, mocking behaviour and an attempt by an individual, perhaps a senior manager or other person in authority, to belittle a victim in the presence of other persons. Inevitably, the victim will feel at risk, distressed and vulnerable, may suffer shock and, in some cases, may require long-standing psychological treatment.

HSE guidance

The HSE in their publication *Violence at Work: A guide for employers* define work-related violence as 'any incident in which a person is abused, threatened or assaulted in circumstances relating to their work'. Clearly, violence may take many forms. The victims of violence, furthermore, may not necessarily be employees but can include customers, members of the public, bystanders and, in some cases, individual managers who may be at risk for a variety of reasons.

Many people who have direct contact with the public, such as shop assistants, bar staff, waiters in restaurants, bank and building society employees, local authority officers and police officers, may face aggressive or violent behaviour. They may be verbally abused, sworn at, threatened with violence or actually assaulted.

Organisations need to consider, therefore, the potential for violence towards their employees and adopt appropriate strategies to protect them while they are at work.

Physical violence

Certain potential victims of physical violence are obvious. These include, in particular, anyone who handles cash or valuable items such as jewellery, namely employees in banks, building societies, money-lending organisations and jewellers' shops. However, other premises are frequently subject to robberies, such as take-away food premises, off-licences, supermarkets and petrol filling stations. This risk has increased dramatically in the last decade due to the extended opening hours operated by many organisations, including 24-hour opening by supermarkets and petrol filling stations.

Traditionally the armed forces and police officers, together with groups such as security personnel, are trained to deal with physical violence. However, other groups, such as teachers, people employed in the delivery and collection of goods, employees in the care industry and, indeed, anyone providing a service, may need such training in order to ensure their safety at work. This aspect may well be one of the outcomes of a risk assessment required under the MHSWR.

Psychological violence

As commercial life becomes more competitive many employees feel threatened and under pressure due to the intensive and often insensitive nature of the management culture in which they work. As organisations strive towards greater financial success, as managers are put under greater pressure to achieve financial and other performance-related objectives, the greater is the potential for the threatening and bullying of employees by their superiors in order to ensure that these objectives are achieved.

This ruthless culture is a common feature of some organisations, resulting in increased sickness absence among staff, the need for counselling and, in some cases, increased accidents.

The legal position

Where there is clear-cut evidence to indicate that employees may be exposed to risks of both physical and psychological violence by virtue of the tasks they undertake, the services they provide and / or the groups of people with whom they may come into contact, employers could well be deemed to be negligent by a civil court if an employee were injured. Clearly, the common duty of care that exists between an employer and an employee must take into account the nature of these risks, and where it could be shown that an organisation failed to show reasonable care towards such persons to whom this duty is owed, then such an organisation could be deemed to be negligent.

The general and specific duties of employers towards their employees under criminal law is well established under Section 2 of the HASAWA. Similar provisions apply in the case of non-employees, eg members of the public, customers, delivery and security personnel.

Where there is a risk to employees (and indeed, in some cases, persons not in his employment), of physical violence in particular, as with a bank or building society offices, jeweller's shop or 24-hour service supermarket or petrol filling station, an employer must be particularly aware of his duties under the MHSWR, particularly with regard to risk assessment.

Assessing the risks

Many factors need consideration as part of the risk assessment process. For instance:

1. What group or groups of employees may be subject to violence?
2. Are there particularly vulnerable jobs or tasks?
3. Does the timing of opening or closing of premises, receipt of deliveries and despatch of goods allow for incidents to be planned?
4. Do employees need to retain large amounts of cash at their workstations, eg supermarket cashiers?
5. How often is cash moved to a safe area and is there a well-controlled system for doing this?
6. Is there a policy for dealing with complaints and difficult customers?
7. Is there evidence of a culture of bullying, aggressive behaviour, victimisation and harassment in the organisation?
8. Are some employees commonly working on their own?
9. Where employees work away from base, is there a system for keeping in touch and, if necessary, providing an escort?
10. Do employees meet clients in their own homes?
11. Are employees provided with separate and safe parking areas?
12. Is access to vulnerable areas, such as bank cashier stations, adequately controlled?
13. Is there a well-established emergency call system in offices where employees may be meeting members of the public and customers?
14. Is closed circuit television installed where employees may be handling cash or dealing with difficult customers?

Prevention and control strategies

1. Staff should be told what is expected of them if there is a robbery, for example:
   (a) how to raise the alarm;
   (b) where to go for safety; and
   (c) that they should not resist or follow violent robbers.
2. There should be clear visibility and adequate lighting so that staff can leave quickly or get help.
3. The build-up of cash in tills should be prevented, and suitable measures should be adopted to move cash safely.
4. Arrangements should be made for staff to have access to a secure location.
5. High-risk entrances, exits and delivery points should be monitored.
6. Buildings should be brightly illuminated and any possible cover for assailants removed.
7. Screens or similar protective devices for areas where staff are most at risk should be provided.
8. Only experienced or less vulnerable staff should be used for high-risk tasks.
9. High-risk jobs should be rotated so that the same person is not always at risk; for particularly high-risk tasks the number of staff should be doubled.
10. Additional staff should be provided for high-risk mobile activities or communication links to base.
11. Personal alarms for high-risk staff should be provided.
12. Signs asking those wearing crash helmets to remove same should be displayed in prominent positions.
13. All staff should receive training in recognising and dealing with violence and the potential for violence.

Further guidance is provided in the HSE publication *Violence at Work: A guide for employers*, in which the HSE recommend a staged approach to effective management of violence.

Helping the victims

The HSE guidance makes the following points. If there is a violent incident involving the workforce an employer will need to respond quickly to avoid any long-term distress to employees. It is essential to plan how any support will be provided. The following should be considered:

(a) *Debriefing* Victims will need to talk through their experience as soon as possible after the event. Remember that verbal abuse can be just as upsetting as a physical attack.

(b) *Time off work* Individuals will react differently and may need differing amounts of time to recover. In some circumstances they may need specialist counselling.
(c) *Legal help* In serious cases legal help may be appropriate.
(d) *Other employees* They may need guidance and/or training to help them to react appropriately.

The Home Office leaflet 'Victims of Crime' gives additional useful advice in the event of an employee suffering an injury or loss or damage from a crime, including how to apply for compensation. The leaflet should be available from libraries, police stations, Citizens' Advice Bureaux and victim support schemes.

Further help

This may be available from local victim support schemes. Your local police station can direct you to your nearest one. Alternatively you can contact them yourself at the addresses below.

In England:

Victim Support, National Office,
Cranmer House, 39 Brixton Road,
London SW9 6DZ Tel: 020 7735 9166

In Scotland:

Victim Support Scotland,
14 Frederick Street, Edinburgh
Tel: 0131 668 2556

Policy on violence at work

For many organisations a policy on violence at work may be necessary to state the organisation's intentions. Such a policy, while relatively short, should be incorporated as a subsection to the statement of health and safety policy.

POST-TRAUMATIC STRESS DISORDER (PTSD)

This is an extremely disabling anxiety disorder that can develop after exposure to some form of ordeal or frightening event in which serious physical harm was sustained or was threatened. Typical work events that can trigger PTSD include violent physical assaults, such as mugging or rape, and being exposed to death and disaster situations, such as fatal accidents or explosions and serious major injury accidents. People who witness certain traumatic events and the survivors of these events are among those at risk of developing the disorder. In some cases, families of victims have also developed PTSD.

Many people suffering from PTSD repeatedly re-experience the particular event in the form of flashbacks, nightmares or frightening thoughts, especially when exposed to situations or objects reminding them of the trauma they suffered. Other symptoms include a total lack of emotion to certain situations, disturbance of sleep, anxiety, depression, irritability and occasional outbursts of anger. Many people suffer intense feelings of guilt. The disorder can arise at any age, including during childhood. Symptoms commonly appear within three months of the frightening event, although in some cases they may start several years later.

Treatment of those suffering from PTSD commonly includes cognitive behavioural therapy, group therapy and exposure therapy. With this therapy the patient slowly and repeatedly relives the frightening ordeal under controlled conditions to help him work through the trauma. Counselling, which enables the patient to talk through his experiences immediately after the event, may reduce some of the symptoms.

PTSD is commonly accompanied by depression, alcohol and other substance abuse and anxiety disorders. On this basis, these other conditions must be treated at the same time if the treatment is to be successful. Patients frequently suffer headaches, chest pain, stomach complaints, immune system disorders, dizziness and discomfort in other parts of the body.

CONCLUSION

This chapter has endeavoured to give a broad overview of the problem of stress at work, a subject which is rarely considered. The most difficult task is getting people to recognise the existence of the stress response within individuals and within the organisation and that their decisions could be stressful for other people. Many managers still adopt the Victorian maxim 'If you can't stand the heat, get out of the kitchen!' Such a response is totally unhelpful to people going through stressful events in their lives, be they associated with the work situation or private life.

SUMMARY

1. Stress is a common feature of most people's lives and the causes of stress are many and varied. It is most commonly associated with changes in people's lives, some of which may be brought about by the organisation.

2. There is a need within organisations for a greater understanding of the stress response and the causes of stress.

3. Stress reduction strategies should be considered at boardroom level and implemented wherever necessary.

4. The costs of stress-related ill health can be substantial in terms of time lost for conditions diagnosed as 'anxiety state', 'depression' and 'nervous breakdown'.
5. Where employees may be exposed to the risk of violence, employers must take appropriate measures to prevent or control these risks.
6. Employers need to pay attention to recent civil law decisions on the subject of stress-induced injury, and to the HSE management standards.

CONCLUSION TO PART 2

Human factors are becoming an increasingly significant feature of health and safety management. The need to recognise the various influences on people at work, in particular those characteristics of an organisation which influence safety-related behaviour, cannot be over-emphasised. In the design of jobs and safe systems of work consideration must be given to the potential for human error, human capabilities and fallibilities and the causes and effects of stress.

We do not plan our working systems in many cases, particularly with regard to the effective use of human resources. The application of ergonomic principles, including safe manual practices, would bring about great reductions in the stress associated with many tasks. Companies who adopt ergonomic principles in their working systems have been able to show reductions in accidents and ill health with the resulting improvements in productivity and profitability.

REFERENCES

Allport, G W (1961) *Pattern and Growth in Personality*, Holt, New York

Bond, M and Kilty, J (1982) *Practical Methods of Dealing with Stress*, University of Surrey, Guildford

Cooper, C L, Cooper, R D and Eaker, L H (1988) *Living with Stress*, Penguin Books, London

Cooper, C L and Marshall, J (1978) *Understanding Executive Stress*, Macmillan, London

Cox, T (1978) *Stress*, Macmillan, London

Health and Safety Executive (1993) *Mental Distress at Work*, HSE Information Centre, Sheffield

Health and Safety Executive (1995) *Stress at Work*, HSE Books, Sudbury

Health and Safety Executive (1998) *Help on Work-Related Stress: A short guide*, HSE Books, Sudbury

Health and Safety Executive (1998) *Violence at work: A guide for employers*, HSE Books, Sudbury

Health and Safety Executive (2005) *Tackling Stress: The management standards approach* [online] www.hse.gov.uk/stress/standards/pdfs/shortguide.pdf (accessed 30 May 2005)
Orlans, V and Shipley, P (1983) *A Survey of Stress Management and Prevention Facilities in a Sample of UK Organisations*, Birkbeck College, University of London
Selye, H (1936) *The Stress of Life*, rev. ed. 1976, McGraw-Hill, New York
Stranks, J (2005) *Stress At Work: Management and prevention*, Elsevier Science & Technology, London

Part 3

Occupational Health

8

Occupational Diseases and Conditions

OCCUPATIONAL HEALTH

Every year people at work contract various forms of occupational disease or condition. Many people die as a result of contracting, for instance, occupational cancer, pneumoconiosis or chemical poisoning. Others may be permanently incapacitated through contracting conditions such as noise-induced hearing loss (occupational deafness), vibration-induced injury or occupational asthma. Occupational dermatitis is the most common form of occupational disease.

Occupational health is a preventive form of medicine that examines, firstly, the relationship of work to health and, secondly, the effects of work on the worker.

Occupational health hazards can be classified as follows.

1. Physical

The effects of exposure to extremes of temperature and humidity, inadequate lighting and ventilation, noise and vibration, dusts, pressure and radiation can result in a range of conditions, such as heat stroke, heat cataract, noise-induced hearing loss, vibration-induced white finger, pneumoconiosis, decompression sickness and radiation sickness.

2. Chemical

The effects of exposure to toxic, corrosive, harmful or irritant solids, liquids, fumes, mists and gases are responsible for various forms of metallic and chemical poisoning, dermatitis and occupational cancers. Clearly the form taken by a substance, eg gas, dust, mist, is significant in its potential for harm.

3. Biological

A range of diseases, such as various forms of human anthrax, leptospirosis, brucellosis, viral hepatitis, legionnaires' disease and aspergillosis (farmer's lung) are caused through exposure to a range of micro-organisms, such as viruses and bacteria. Certain diseases, such as brucellosis and anthrax, are transmissible from animals to humans (zoonoses).

4. Ergonomic (work related)

The effects on people of poorly designed working layouts and operator workstations, together with excessive and repetitive movements of joints, can lead to visual and postural fatigue, physical and mental stress and a range of conditions. These include, for instance, writer's cramp, various 'beat disorders' (beat knee, beat wrist and beat elbow) and that group of conditions known as the 'work-related upper limb disorders' or repetitive strain injury. This last-mentioned group includes conditions such as tenosynovitis and carpal tunnel syndrome. The potential for stress-induced injury must also be considered. (See Chapter 6.)

Various conditions arising from manual handling operations, such as prolapsed intervertebral discs, hernia and ligamental strain also come into this category.

Factors for consideration

Any strategy designed to prevent or reduce the risk of employees contracting various forms of ill health arising from work activities should take the following factors into account:

(a) measures to prevent or reduce the risk of occupational disease, including health surveillance procedures;
(b) systems for the identification, measurement, evaluation and control of occupational health risks in the working environment;
(c) welfare amenity provisions including sanitation arrangements, hand-washing and shower facilities, clothing storage, taking of meals and the provision of pure drinking water;

(d) first aid arrangements, including the training of first aid staff, and the provision of emergency services in the event of serious injury;
(e) the ergonomic aspects of certain jobs;
(f) the selection, provision, assessment of suitability, maintenance and use of personal protective equipment; and
(g) measures to ensure the provision of information, instruction, training and constant supervision for all persons exposed to health risks.

Practitioners in occupational health

Various specialists may be involved in occupational health practice. These include occupational physicians, occupational hygienists, occupational health nurses and health and safety practitioners. Each of these groups have a specific contribution to make in preventing ill health arising from workplace operations and activities.

OCCUPATIONAL DISEASES

This is a group of diseases and conditions contracted as a result of a particular employment. As such, they may be 'prescribed' under the Social Security Act 1975, in which case benefit is payable to those suffering from them, and 'reportable' under RIDDOR.

Prescribed diseases

Prescribed diseases are listed in Schedule 1 of the Social Security (Industrial Injuries) (Prescribed Diseases) Regulations 1985. Each prescribed disease is related to a specific occupation. A disease is 'prescribed' if:

(a) it ought to be treated, having regard to its causes and incidence and other relevant considerations, as a risk of occupation and not as a risk common to all persons; and
(b) it is such that, in the absence of special circumstances, the attribution of particular cases to the nature of the employment can be established with reasonable certainty.

(Social Security Act 1975, Section 76(2))

Reportable diseases

These are diseases which are reportable to the enforcement agency, eg HSE, local authority, under RIDDOR (Regulation 5). In this case, a wide range of diseases is listed and qualified by a particular work activity, together with a common description of the disease. Typical examples include poisoning by carbon disulphide, methyl bromide, etc; skin diseases, such as folliculitis; occupational asthmas; pneumoconiosis; and certain infections, such as hepatitis. Such diseases must be reported to the enforcement agency on Form 2508A.

Principal causes and effects

RIDDOR classifies occupational diseases on the basis of causation and the work activity. The Social Security (Industrial Injuries) (Prescribed Diseases) Regulations classify in a similar way on the basis of occupation.

Occupational diseases and conditions are classified in the following ways.

Conditions due to physical agents

1. Temperature – heat stroke, heat cataracts.
2. Lighting – miner's nystagmus.
3. Radiation – radiation sickness, arc eye, burns.
4. Noise – noise-induced hearing loss.
5. Vibration – vibration-induced white finger.
6. Pressure – decompression sickness.
7. Dust – pneumoconiosis, including silicosis, coal worker's pneumoconiosis, occupational asthma, occupational cancers.
8. Repetitive movements – writer's cramp.
9. Manual work – beat elbow, beat knee, beat hand.

Conditions due to biological agents

1. Contact with infected animals – anthrax, brucellosis, glanders.
2. Contact with blood or blood products – viral hepatitis.
3. Vegetable-borne infections – farmer's lung (aspergillosis).
4. Contact with rodents – leptospirosis.

Conditions due to chemical agents

1. Poisonings – the use or handling of, or exposure to, the fumes, dust or vapour of a wide range of chemical substances, including lead, manganese, phosphorus, arsenic and mercury.
2. Occupational cancers – associated with exposure to tar, pitch, bitumen, mineral oils, aromatic amines, such as alpha-naphthylamine and beta-naphthylamine, and vinyl chloride monomer (VCM).
3. Specific damage to organs and systems – the use or handling of, or exposure to the fumes of, or vapours containing solvents, such as carbon tetrachloride, trichloromethane (chloroform) and chloromethane (methyl chloride).

Conditions due to work activities

1. Job movements – writer's cramp, 'beat' conditions eg beat hand, beat elbow, beat knee, traumatic inflammation of the tendons or associated tendon sheaths of the hand or forearm ie tenosynovitis, work-related upper limb disorders (repetitive strain injury).
2. Friction and pressure – bursitis, cellulitis.

THE EMPLOYMENT MEDICAL ADVISORY SERVICE (EMAS)

The EMAS is fundamentally the Medical Services Division of the HSE. It comprises a national network of around 150 doctors and nurses accountable to a number of Senior EMAs and the Director of Medical Services. Approximately 22,000 statutory medical examinations are undertaken every year, together with 90,000 such examinations carried out by company occupational physicians who are specified as 'appointed doctors' under the system.

Under these statutory requirements:

(a) the employer is officially notified of the fitness of the worker to undertake employment;
(b) the employee has a duty to undergo examinations;
(c) the employer cannot lawfully continue to employ any worker who is found to be unfit;
(d) the employee must be removed from that particular job for the prescribed period and transferred to alternative work if it is available; and
(e) the outcome of the medical examination must be recorded in a specific health register by the employer.

OCCUPATIONAL HEALTH SCHEMES

Many organisations operate occupational health schemes for their employees. Generally, these schemes are managed and implemented by occupational health nurses with, perhaps, an occupational physician available to provide a medical input when required, and embrace some or all of the main areas of occupational health practice. In certain cases, a scheme may be directed at complying with specific health surveillance requirements laid down in regulations, such as the MHSWR and the COSHH Regulations.

Occupational health practice incorporates the following aspects:

1. **Placing people in suitable work**
   This entails the assessment of current and mental and physical capability and identification of pre-existing ill-health conditions. It generally

takes the form of pre-employment medical examinations and/or pre-employment health screening.

2. **Health surveillance**
   The provision of health surveillance for certain employees may be one of the outcomes of a risk assessment under the MHSWR, COSHH Regulations or Control of Noise at Work Regulations. It entails specific health examinations at a predetermined frequency for:
   (a) those at risk of developing further ill health or disability, such as employees who may be occasionally exposed to excessive noise levels or hazardous dusts; and
   (b) those actually or potentially at risk by virtue of the type of work they undertake during their employment, such as radiation workers.
3. **Providing a treatment service**
   This involves the efficient and speedy treatment of injuries, acute poisonings and minor ailments at work. This service is important in terms of keeping people at work, thereby reducing lost time associated with attendance at local doctors' surgeries and accident and emergency departments.
4. **Primary and secondary monitoring**
   Primary monitoring is principally concerned with the clinical observation of sick persons who may seek treatment or advice on their condition. *Secondary monitoring*, on the other hand, is directed at controlling the hazards to health which have already been recognised, for example, regular audiometry for employees exposed to noise.
5. **Avoiding potential risks**
   This is an important aspect of occupational health practice with the principal emphasis on prevention, in preference to treatment, for a known condition. This may entail, for example, making recommendations with respect to the substitution of certain hazardous substances with less hazardous substances.
6. **Supervision of vulnerable groups**
   Vulnerable workers include young persons, pregnant women, the aged, the disabled and persons who may have had long periods of sickness absence, perhaps as a result of surgery. New and expectant mothers and young persons are two groups singled out in the MHSWR with respect to risk assessment requirements. Regular health examinations to assess continuing fitness for work may be necessary for people in these groups.
7. **Monitoring for evidence of non-occupational disease**
   This is a form of routine monitoring of employees not exposed to health risks with the principal objective of controlling diseases and conditions prevalent in certain communities, such as mining, with a view to their eventual eradication.

8. **Counselling**
   Counselling employees on a range of health-related issues, and on personal, social and emotional problems, is an important feature of an occupational health service.
9. **Health education**
   This is primarily concerned with the education of employees towards a healthier lifestyle. It can also include the training of management and employees in various areas of health and safety at work, in healthy working techniques and in the avoidance of health hazards, such as those arising from manual handling and the use of hazardous substances.
10. **First aid and emergency services**
    Occupational health services commonly supervise first aid arrangements, train first aid staff and prepare specific aspects of contingency arrangements in the event of fire or other disaster situations, such as a major chemical spillage or an explosion.
11. **Occupational hygiene**
    Occupational hygiene is an area of occupational health practice concerned with the identification, measurement, evaluation and control of contaminants, and other physical phenomena, such as noise and radiation, which could have adverse effects on the health of people exposed to such contaminants. Occupational health services frequently employ occupational hygienists to concentrate specifically on this area of risk.
12. **Environmental control**
    This area is mainly concerned with ensuring compliance with environmental protection legislation, such as the Environmental Protection Act, and covers measures for the prevention and control, in particular, of airborne contaminants, such as dusts, gases and fumes, together with noise, which could be a nuisance to the inhabitants of the neighbourhood.
13. **Health records**
    The completion and maintenance of health records required under certain regulations, such as the COSHH Regulations, Noise at Work Regulations and Control of Lead at Work Regulations, is commonly undertaken by occupational health services. These records may further be necessary on an internal basis for ensuring health surveillance procedures are maintained and may feature in epidemiological studies of certain groups of employees.
14. **Liaison**
    It is standard practice for occupational health practitioners to liaise with enforcement agencies, in particular the Employment Medical Advisory Service (EMAS) of the HSE.

BENEFITS OF OCCUPATIONAL HEALTH SERVICES

One of the principal functions of an occupational health service is that of keeping people at work through the provision of on-site treatment, counselling and advice with respect to health conditions. Sickness absence, and the cost of same, is markedly reduced, together with the lost time costs of injured employees, managers and others following accidents at work.

In addition to the improved morale and commitment such a caring service creates among employees, many forms of ill health, not necessarily arising from work, are diagnosed at an early stage, permitting prompt treatment and an early return to work by those affected. Procedures such as audiometry and vision screen of employees enable early signs of deafness and visual defects to be identified where, in many cases, an employee would not normally have sought help from a doctor.

The provision of advice to managers with respect to vulnerable groups, health education needs, emergency procedures and dealing with enforcement officers is a further valuable component of such a service.

SICKNESS ABSENCE CONTROL

The cost of sickness absence to all forms of commercial undertaking can be substantial if procedures are not established and followed. Over 200 million man days are lost each year due to absence from work, a large proportion of which is attributed to sickness. With approximately 20 million people employed in the United Kingdom, this amounts to between 10 and 11 days' absence per person per year – ie an absence rate of 4.6 per cent. Absence can be associated with frequent periods of short-term sickness and prolonged periods of sickness absence. Failure to produce sick notes and the taking of frequent short unconnected periods of absence can, in certain cases, result in an employee being dismissed. Much will depend upon that employee's terms of contract, the statement of terms and conditions, features of the company sick pay and absence scheme and the specific action taken by the employee. Non-compliance with the company scheme may result in breach of contract with subsequent withholding of his pay. The company, on investigating the reasons for frequent periods of short-term sickness absence, may decide the employee is guilty of misconduct and, on the basis of a national agreement, take disciplinary action. The disciplinary action could be a verbal or written warning, or alternatively, dismissal.

Long-term sickness absence hinges around 'capability' to work, rather than misconduct. Capability to work is assessed by reference to skill, aptitude, health or any other physical or mental quality. Where the employer makes a decision to terminate employment on the basis of incapability to work, he should first consider the potential for his being taken to a tribunal on the grounds of unfair dismissal. The following questions should, therefore, be asked:

1. Were offers of alternative employment considered?
2. Was the employee consulted sufficiently about his problems, future employment prospects and the possibility of dismissal?
3. Was the latest period of sickness absence prior to receiving the warning of the possibility of dismissal? Was a further warning appropriate in this case?
4. Has there been a recent opportunity for the employee to comment on his health problem and ability to work? Was an independent medical opinion sought?
5. Has the employer sought medical advice over this situation?
6. Has the employer investigated all relevant matters prior to the dismissal decision?
7. How long, apart from the period of illness, would the employment be likely to last?
8. Can the employer wait any longer for the employee to return to work? (A balance must be achieved between the position of the employee, the interests of the company and the need to be fair.)
9. How important is it for this employee to be replaced?
10. Has the employee been consulted on the final step in the procedure?

Once these questions have been answered satisfactorily, a decision may need to be taken as to whether or not employment should be terminated. The four questions below should then be considered:

1. Were the terms of the contract of employment, including sick pay provisions, fulfilled?
2. What is the nature of the employment, with particular reference to a key position?
3. What is the nature of the illness/injury? How long has it continued and what are the prospects of recovery?
4. What is the total period of employment?

At completion of this exercise, the employee must be interviewed, informed of management's decision and the result confirmed in writing.

In the event of an appeal to an employment appeals tribunal, the tribunal would be seeking evidence of the following having been carried out by the employer:

1. A fair review of the employee's attendance record and the reasons for periods of absence.

2. Evidence of the employee having received appropriate warnings after he had been given an opportunity to make representations.

Where there has been no improvement in the attendance record, it is likely that, in most cases, the employer would be justified in treating the persistent absences as sufficient reason for dismissing the employee.

It will be seen that sickness absence control hinges around a clearly established procedure linked to the allocation of sick pay.

SICKNESS ABSENCE PROCEDURE

A typical company sickness reporting and pay procedure should take the form shown below.

IN ORDER TO QUALIFY FOR SICK PAY ALL STAFF MUST

1. Notify their supervisor/foreman by 10.00 am (or at least two hours prior to the commencement of a shift) on the first day of absence, giving an indication as to the expected length of absence.
2. Where sick between one and three days, employees must complete the company self-certification form; if not submitted, payment can be withheld.
3. Where sick between four to seven days, employees must submit a DSS self-certification form; if not submitted, payment can be withheld.
4. If sick for more than seven days, a doctor's certificate must be submitted; if not submitted, payment can be withheld.
5. On return to work, the employee must notify his supervisor of his intention to return by 2.00 pm on the day prior to return, or for night-shift workers, by 12 noon of the day of the shift.

Note:

1. In cases 2 and 3 above, submission of the self-certification form should be made on the day of return to work.
2. A doctor's certificate should be submitted within at least 10 days of the commencement of the period of absence. If the certificate is not dated from the first day of absence, the employee must complete the self-certification form.

SICKNESS ABSENCE MONITORING

All supervisors / foremen should maintain a day-to-day record of employees' attendance at work. This can be done on a monthly basis whereby, on each working day, a record is made of those employees who are present, those who are absent for a particular reason (eg annual leave, rest day, etc), and those who are absent through sickness or unexplained reasons. On this basis it is relatively simple to identify patterns of short-term sickness absence among the workforce and to instigate early investigatory processes to determine the causes of such absences. Employees should be reminded regularly of the need to follow the sickness absence procedure above, and disciplinary action should be taken against persistent defaulters.

The procedure should ensure that employers act as follows:

1. They tackle the problem face to face.
2. They observe and record the problem.
3. They find out the facts.
4. They agree a plan of action.

The ultimate objectives of a sickness absence control system are to:

(a) recognise and measure absenteeism associated with ill health;
(b) identify causes of such absence in the workplace;
(c) deal fairly with employees who are frequently absent due to ill health;
(d) set up procedures to discourage absenteeism; and
(e) recognise when the company needs specialist help in dealing with the problem.

SMOKING AT WORK

The relationship of cigarette smoking in particular to various forms of cancer is well established. Smoking has a direct effect on people in terms of reduced lung function and an increased potential for lung conditions, such as bronchitis. The problem has been identified mainly in poorly ventilated areas and among employees who may suffer some form of respiratory complaint, such as asthma or bronchitis. Other people may complain of soreness of the eyes, headaches and stuffiness.

Measures to reduce the risk to health associated with smoking are incorporated in the Health Act 2006. This legislation covers all premises which are wholly or substantially enclosed and used as a place of work by more than one person. Fundamentally, it ensures that all workers, regardless of their place of work, are protected from the risks to health of exposure to tobacco smoke and guaranteed the right to smoke-free air.

Premises are considered substantially enclosed if they have a ceiling or roof and the openings in the walls are less than half the total area of the walls. Smoking is permitted in shelters which are not substantially enclosed.

Under the Act, employers, owners of premises and managers must ensure their premises are smoke-free. They must display the approved 'No Smoking' signs at each public entrance to the premises and in a position which is prominently visible to persons entering the premises. Similarly, vehicles used as a workplace by more than one person, irrespective of whether they are all in the vehicle at the same time, must be smoke-free at all times.

Local authorities have a duty to enforce the requirements of the Act and must identify enforcement officers, eg environmental health officers, who are authorised to take a range of enforcement actions, such as the issue of penalty notices. The final penalty amount is decided by a court sooner than the local authority. The fixed penalties and maximum fines levied by a court are:

- *Failure to display minimum 'No Smoking' signs.*
  A fine on the owner or manager of up to £1,000 or issue of a fixed penalty notice for £200.
- *Smoking in a smoke-free place.*
  A fine on the person of up to £200 or issue of a fixed penalty notice for £50.
- *Failure to prevent smoking in a smoke-free place.*
  A fine on the owner or manager of up to £2,500.

Drugs in the workplace

Many people see drug taking as the panacea for stress, relying on tranquillisers to reduce anxiety and amphetamines (pep pills) to counter fatigue. This form of drug taking represents a major risk to health, particularly if the individual consumes alcohol.

The possession of a range of drugs is a criminal offence. Evidence of drug taking or 'pushing' in the workplace should be reported to a senior manager. Where appropriate, the advice of the police should be obtained. A programme to assist identified drug users to break the habit may need to be considered and put into operation. A range of national agencies are available to assist in such situations.

Alcohol at work

The consumption of alcohol at work may be one of the outcomes of stress in certain cases. Alcoholism is a true addiction, and the alcoholic must be encouraged to obtain medical help and advice.

The repeated consumption of strong spirits, especially on an empty stomach, can lead to chronic gastritis and possible inflammation of the intestines that interferes with the absorption of food substances, notably those in the vitamin B group. This, in turn, damages the nerve cells causing alcoholic neuritis, injury to the brain cells leading to certain forms of insanity and, in some cases, cirrhosis of the liver.

A healthy lifestyle

Regular exercise, such as swimming, taking sufficient rest and paying attention to the food one eats are all features of a healthy lifestyle.

SUMMARY

1. Poor standards of health control in the workplace can result in employees contracting occupational disease and conditions.
2. The differences between 'prescribed' and 'reportable' diseases should be recognised and understood.
3. The majority of occupational diseases are associated with physical, chemical and biological agents present at work.
4. Staff of the EMAS have powers to take action against employers and occupiers of premises in certain cases where there may be a risk to the health of workers, including the compulsory medical examination of workers.
5. An organisation's sickness absence costs can be substantial if no procedure for monitoring and regulating them is operated.

REFERENCES

Atherley, G R C (1978) *Occupational Health and Safety Concepts*, Applied Science Publishers Ltd, London

Department of Health and Social Security (1980) *Social Security (Industrial Industries) (Prescribed Diseases) Regulations 1985 (SI 1985 No. 967) and subsequent Amendment Regulations 1986 & 1987*, HMSO, London

Department of Health and Social Security (1983) *Notes on the Diagnosis of Occupational Diseases*, HMSO, London

Health and Safety Executive (1981) *Health Surveillance by Routine Procedures, Guidance Note MS 18*, HMSO, London

Health and Safety Executive (1988) *Passive Smoking at Work*, HSE Information Centre, Sheffield

Reporting of Injuries, Diseases and Dangerous Occurrences Regulations 1985 (SI 1985 No 2023), HMSO, London

Stranks, J (1995) *Occupational Health and Hygiene*, Pitman, London

9

First Aid

The provision of facilities for first aid treatment in workplaces is a basic legal requirement under the FA, OSRPA and other principal protective legislation.

Legal considerations apart, however, there is clearly a case for as many people as possible being trained, irrespective of their functions, in basic first aid procedures. Such knowledge could be instrumental in saving the life of a person whatever the circumstances. This philosophy has been promoted by many organisations, including the national first aid organisations, following recent disasters involving both workers and members of the public.

Current legal requirements are covered by the Health and Safety (First Aid) Regulations 1981, made under the HASAWA, an ACOP and guidance issued by the HSE.

HEALTH AND SAFETY (FIRST AID) REGULATIONS 1981

Under these regulations, 'first aid' is defined as meaning:

(a) in cases where a person will need help from a medical practitioner or nurse, such treatment necessary to preserve life and minimise the consequences of injury and illness until such help is obtained; and
(b) treatment of minor injuries which would otherwise receive no treatment or which do not need treatment by a medical practitioner or nurse.

There are three general duties under the regulations, namely:

(a) the duty of the employer to provide first aid arrangements;
(b) the duty of the employer to inform his employees of these arrangements; and
(c) the duty of the self-employed person to provide first aid equipment.

These duties are expanded further in the ACOP and guidance notes issued with the regulations. There are four specific criteria which the employer must take into account in deciding what are adequate and appropriate arrangements for first aid in the workplace, namely:

(a) the number of employees;
(b) the nature of the undertaking;
(c) the size of the establishment, and distribution of employees; and
(d) the location of the establishment and of the employees' places of work.

An adequate number of suitable persons able to give first aid must be provided, and these persons must be adequately trained and hold an appropriate first aid qualification approved by the Health and Safety Executive. In certain circumstances, eg high-risk activities, additional training may be necessary.

Note: It should be stressed, however, that cover must be adequate at all times when people are at work and there is clearly a case, therefore, for having more trained first-aiders available than shown above in order to cover for holidays, sickness, shift working, etc.

In situations where it is not necessary to appoint a first-aider, an appointed person must be designated who can take charge of situations, ie call a doctor and/or ambulance, in the event of serious injury or major illness.

Many employees work away from a main location. In these cases the employer has the responsibility to make adequate and appropriate first aid facilities available for these staff. Furthermore, where employees work alone or in small groups, where the work involves travelling long distances, or where employees may be using potentially dangerous equipment or machinery, small travelling first aid kits must be provided.

Contents of first aid boxes and kits are specified in the regulations. They must be checked and replenished as necessary.

First aid rooms

The code of practice provides that an employer should generally provide a suitably equipped and staffed first aid room only where 400 or more employees are at work. However, in the case of:

(a) establishments with special hazards;
(b) construction sites with more than 250 persons at work; and

(c) when access to casualty centres, or emergency facilities, is difficult, eg owing to distance or inadequacy of transport facilities;

a first aid room should be provided. Any first aid room shall be easily accessible to stretchers and to any other equipment needed to convey patients to and from the room, and be signposted in accordance with the Health and Safety (Safety Signs and Signals) Regulations 1996.

The role of first-aiders

The ACOP places special emphasis on the need for employers to avail themselves of first-aiders trained in specific techniques required by their undertaking. In certain cases, for instance where there may be a risk of cyanide poisoning or there may be a need for oxygen for resuscitation purposes, training should be undertaken by organisations approved by the HSE.

The provision of trained first-aiders, furthermore, is related to specific circumstances, as in the case of shift work, where there must be adequate coverage for each shift. In the case of low-risk establishments eg offices, shops, there need normally be no first-aider where fewer than 150 employees are at work, with at least one first-aider for 150 employees or more. For establishments with greater risk, eg factories, farms, the employer should provide one first-aider per 50–150 employees, and an additional first-aider where there are more than 150 employees. Where there are fewer than 50 employees, an employer must in any case provide an appointed person.

First aid equipment

Where first aid boxes form part of an establishment's first aid provision, they should contain only those items which a first aider has been trained to use. Sufficient quantities of each item should always be available in every first aid box or container. In most cases, these will be:

(a) one guidance card;
(b) twenty individually wrapped sterile adhesive dressings (assorted sizes) appropriate to the work environment (which may be detectable for the food and catering industries);
(c) two sterile eye pads with attachment;
(d) six individually wrapped triangular bandages;
(e) six safety pins;
(f) six medium-sized individually wrapped sterile unmedicated wound dressings (approx. 10 cm × 8 cm);
(g) two large sterile individually wrapped unmedicated wound dressings (approx. 13 cm × 9 cm); and

(h) three large sterile individually wrapped unmedicated wound dressings (approx. 28 cm × 17.5 cm).

Where mains tap water is not available for eye irrigation, sterile water or sterile normal saline (0.9 per cent) in sealed disposable containers should be provided. Each container should hold at least 300 ml and should not be reused once the sterile seal is broken. At least 900 ml should be provided. Eye baths / eye cups / refillable containers should not be used for eye irrigation.

Where an employee has received additional training in the treatment of specific hazards which require the use of special antidotes or special equipment, these may be stored near the hazard area or may be kept in the first aid box.

Travelling first aid kits

In the case of employees who regularly work away from their employer's establishment in isolated locations, or where they are involved in travelling long distances in remote areas from which access to accident and emergency facilities may be difficult, it may be necessary for first aid equipment to be carried by, or made available to, employees where potentially dangerous tools or machinery are used. The equipment should be suitable for the numbers involved and the potential hazards to which employees are exposed.

The contents of travelling first aid kits should be appropriate for the circumstances in which they are used. At least the following should be included:

(a) card giving the general first aid guidance;
(b) six individually wrapped sterile adhesive dressings;
(c) one large sterile unmedicated dressing;
(d) two triangular bandages;
(e) two safety pins; and
(f) individually wrapped moist cleaning wipes.

HSE GUIDANCE – FIRST AID AT WORK

HSE leaflet INDG347, *Basic Advice on First Aid at Work*, makes the following recommendations on what to do in an emergency, based on the mnemonic, A – B – C (airway–breathing–circulation):

What to do in an emergency

Priorities

- Assess the situation – do not put yourself in danger.
- Make the area safe.
- Assess all casualties and attend first to any *unconscious* casualties.
- *Send for help – do not delay.*
- Follow the advice given below.

Check for consciousness

If there is no response to gentle shaking of the shoulders and shouting, the casualty may be unconscious. The priority is then to check the **A**irway, **B**reathing and **C**irculation. This is the **ABC** of resuscitation.

A – Airway

- Place one hand on the casualty's forehead and gently tilt the head back.
- Remove any obvious obstruction from the casualty's mouth.
- Lift the chin with two fingertips.

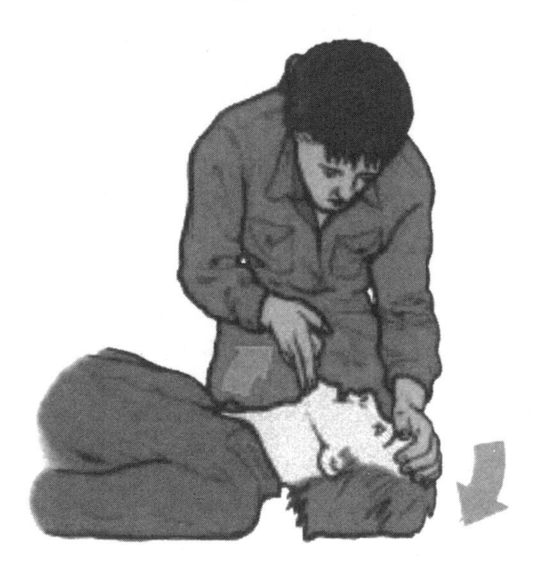

Figure 9.1 Opening the airway

B – Breathing

Look along the chest, listen and feel at the mouth, for signs of normal breathing, for no more than 10 seconds. If the casualty is breathing:

- Place in recovery position and ensure the airway remains open.
- Send for help and monitor the casualty until help arrives.

Figure 9.2 The recovery position

If the casualty is *not* breathing:

- Send for help.
- Keep the airway open by maintaining the head tilt and chin lift.
- Pinch the casualty's nose closed and allow the mouth to remain open.
- Take a full breath and place your mouth around the casualty's mouth, making a good seal.
- Blow slowly into the mouth until the chest rises.
- Remove your mouth from the casualty and let the chest fall fully.
- Give a second slow breath, then look for signs of circulation (see below).
- *If signs of circulation are present,* continue breathing for the casualty and recheck for signs of circulation about every 10 breaths.
- If the casualty starts to breathe but remains unconscious, put him in the recovery position, ensure the airway remains open and monitor until help arrives.

C – Circulation

Look, listen and feel for normal breathing, coughing or movement by the casualty., for no more than 10 seconds. *If there are no signs of circulation, or you are at all unsure, immediately start chest compressions:*

- Lean over the casualty, and with straight arms, press vertically down 4–5 cm on the breastbone, then release the pressure.
- Give 15 rapid chest compressions (a rate of about 100 per minute) followed by two breaths.
- Continue alternating 15 chest compressions with two breaths until help arrives or the casualty shows signs of recovery.

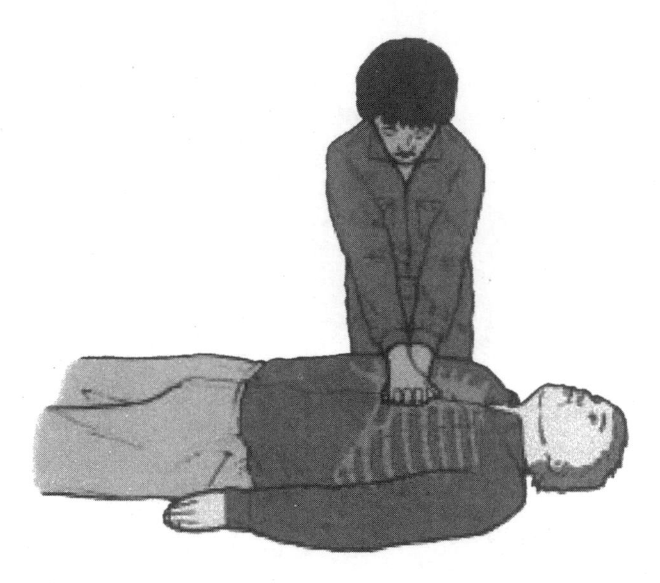

Figure 9.3 Chest compressions

Severe bleeding

- Apply direct pressure to the wound.
- Raise and support the injured part (unless broken).
- Apply a dressing and bandage firmly in place.

Broken bones and spinal injuries

If a broken bone or spinal injury is suspected, *obtain expert help. Do not move casualty* unless he is in immediate danger.

Burns

Burns can be serious so if in doubt, *seek medical help*. Cool the part of the body affected with cold water until pain is relieved. Thorough cooling may take 10 minutes or more, but this must not delay taking the casualty to hospital.

Certain chemicals may seriously irritate or damage the skin. Avoid contaminating yourself with the chemical. Treat in the same way as for other burns but flood the affected area with water for 20 minutes. Continue treatment even on the way to hospital, if necessary. Remove any contaminated clothing that is not stuck to the skin.

Eye injuries

All eye injuries are potentially serious. If there is something in the eye, wash out the eye with clean water or sterile fluid from a sealed container, to remove loose material. Do not attempt to remove anything that is embedded in the eye.

If chemicals are involved, flush the eye with water or sterile fluid for at least 10 minutes while gently holding the eyelids open. Ask the casualty to hold a pad over the injured eye and send him to hospital.

Record keeping

It is good practice to record in a book any incidents involving injuries or illness that have been attended. Include the following information in your entry:

- Date, time and place of the incident.
- Name and job of the injured or ill person.
- Details of injury / illness and any first aid given.
- What happened to the casualty immediately afterwards (for example, went back to work, went home, went to hospital).
- Name and signature of the person dealing with the incident.

This information can help identify accident trends and possible areas for improvement in the control of health and safety risks.

PRINCIPLES OF FIRST AID

The aims of first aid treatment are to sustain life, to prevent deterioration in an existing condition, and to promote recovery.

Principal aspects of such treatment are resuscitation (restoration of breathing), control of bleeding and the prevention of collapse. Reference to the 'general guidance' which must be shown in first aid boxes, clearly reinforces these principal aspects.

All staff, irrespective of whether they are designated first-aiders, should be familiar with the resuscitation procedure shown in Figure 9.1 on page 198. The procedures, in card form, are available from RoSPA.

THE ROLE OF FIRST-AIDERS

A first-aider is a person who has received training and who holds a current first aid certificate from an organisation or employer whose training and qualifications for first-aiders are approved by the HSE.

The definition of 'first aid' in the regulations, or the alternative definition below, are a useful guide in deciding on the first aid arrangements necessary for a particular enterprise. First aid is:

> 'the skilled application of accepted principles of treatment on the occurrence of an accident or in the case of sudden illness, using facilities and materials available at the time'.

The identification of first aid needs, in terms of the number of, and degree of training for first-aiders, provision of facilities and the arrangements for ensuring 100 per cent first aid provision at all times, is a legal requirement. Employers must identify such needs and review them on a regular basis.

SUMMARY

1. The Health and Safety (First Aid) Regulations 1981, together with the ACOP and Guidance Notes, provide guidance for employers on the arrangements necessary for first aid in the workplace.
2. Specific provision must be made for high-risk locations, people who work away from the main location and the giving of information to staff.
3. All staff should be aware of the principles of first aid and basic first aid procedures.
4. The skilled application of first aid procedures, such as restoration of breathing or control of bleeding, has saved many people's lives when injured at work.
5. Procedures for first aid treatment, including management responsibilities for ensuring adequate provision of same, should be clearly identified in the company statement of health and safety policy.

REFERENCES

Health and Safety Commission (1990) *Approved Code of Practice: Health and Safety (First Aid) Regulations 1981*, HMSO, London

Health and Safety Executive (1992) *First Aid Needs in Your Workplace*, HSE Information Centre, Sheffield

Health and Safety Executive (2002) *Basic Advice on First Aid at Work (INDG347)*, HSE Books, Sudbury

Health and Safety (First Aid) Regulations 1981 (SI 1981 No 917), HMSO, London

St John Ambulance Association and Brigade (1972) *First Aid Manual*, SJAB, London

10

Dangerous Substances

Dangerous substances are used in a wide range of industrial and commercial activities and, over the years, many workers have contracted occupational diseases through exposure to such substances. Typical examples include occupational dermatitis (non-infective dermatitis), chemical poisonings (eg by derivatives of arsenic, phosphorus and lead), occupational cancers and the group of diseases known as the pneumoconioses (eg asbestosis, siderosis, coal worker's pneumoconiosis and silicosis). The potential for contracting an occupational disease will vary according to the potential for harm of the substance concerned, its form (eg solid, liquid, gas, dust, etc), the route of entry into the body, the precautions taken (both physically by the employer and personally by the worker), the dose received on a short-term, medium-term and long-term basis, and the degree of susceptibility to the substance of the individual.

Other substances, while not directly entering the body, can be dangerous – eg flammable, explosive, cryogenic substances, contact with which can result in burns, physical damage such as major injuries, and skin irritation.

CHEMICALS (HAZARD INFORMATION AND PACKAGING FOR SUPPLY) (CHIP) REGULATIONS 2002

These regulations implement a series of European Community Directives on the classification, packaging and labelling of dangerous substances and preparations. The objective of the CHIP Regulations is to help protect people and the environment from the ill-effects of chemicals. Under the regulations, suppliers are required to:

(a) identify the hazards (or dangers) of the chemicals they supply;
(b) give information about the hazards to the people who are supplied; and
(c) package the chemicals safely.

A *supplier* is someone who supplies chemicals as part of a transaction, eg manufacturers, importers and distributors.

Supply requirements

The supply requirements implement a series of directives adopted by the member states of the European Community (EC) on classification, packaging and labelling of dangerous substances and preparations. (*Substances* are pure chemicals, like ethanol or water, and *preparations* are mixtures of chemicals, like paints or gin. The term *chemical* covers substances and preparations.)

Scope of the CHIP Regulations

This is one of the largest packages of regulations derived from international agreements. It is relevant to all chemicals, from the most complicated and esoteric to household commodities like paint or bleach. The CHIP Regulations, developed to educate companies to take the right precautions, concern anyone who supplies or consigns dangerous chemicals in Great Britain or offshore. This includes not only large companies but also small manufacturers who concentrate on making one or more common chemicals or who retail chemicals to the public.

The CHIP Regulations are also the foundation for other health and safety and environmental provisions. For example, substances classified under the CHIP Regulations as being harmful to health fall within the scope of the COSHH Regulations. The safety data sheets required by the CHIP Regulations are a major source of information for those who have to prepare COSHH assessments.

Certain substances identified as dangerous by the regulations fall within the scope of the Control of Major Accident Hazards (COMAH) Regulations 1999.

The supporting guidance for the CHIP Regulations consists of:

(a) the Approved Supply List: Information Approved for the Classification, Packaging and Labelling of Substances and Preparations Dangerous for Supply;
(b) the Approved Guide to the Classification and Labelling of Substances and Preparations Dangerous for Supply; and
(c) the Approved Code of Practice on Safety Data Sheets for Substances and Preparations Dangerous for Supply.

Classification requirements

The fundamental requirement is for a supplier to decide if the chemical is dangerous or not. If it is dangerous, then it must be classified, and it is an offence to supply a chemical before this classification is complete.

The first step in classification is to put the chemical into a category of danger. There are three main categories of dangerous chemicals, which are further sub-divided:

(a) substances and preparations dangerous because of their physical or chemical properties:
   (i) explosive;
   (ii) oxidising;
   (iii) extremely flammable;
   (iv) highly flammable;
   (v) flammable;

(b) substances and preparations dangerous because of their health effects:
   (i) very toxic;
   (ii) toxic;
   (iii) harmful;
   (iv) corrosive;
   (v) irritant;
   (vi) sensitising;
   (vii) carcinogenic (substances which cause cancer);
   (viii) mutagenic (substances which cause inherited changes);
   (ix) toxic for reproduction (substances which cause harm to the unborn);

(c) substances dangerous for the environment.

Although suppliers are responsible for classification, they do not have to undertake the classification themselves. However, the supplier must ensure that the classification has been carried out by a competent person.

Some 1,400 of the more common dangerous chemicals have been classified – details are given in the Approved Supply List. Each substance is identified by its chemical name or its unique international number, and the classification and other information can be read directly from the list.

Where a substance is not on the Approved Supply List, then suppliers have to classify the substances themselves on the basis of the properties described in the regulations. The competent person must find and assemble the available information on the chemical, but it will not be necessary to carry out any new tests. Once the data has been assembled, it can be compared with the criteria in the Approved Guide to the Classification and Labelling of Substances and Preparations Dangerous for Supply. The competent person must then decide whether the substance is dangerous and what classification (if any) is appropriate.

Preparations must be classified in the same categories, except for the classification for environmental effects, which is not required. Since one of the objectives of the CHIP Regulations is to avoid unnecessary animal testing, there is a method for calculating a preparation's category of danger which avoids the practice altogether.

Safety data sheets

One of the important requirements is that safety data sheets be provided for dangerous chemicals which are supplied for work. A safety data sheet must be provided by the supplier to the recipient, and must contain information about the chemical to enable the recipient to take the right precautions. The ACOP: *Safety Data Sheets for Substances and Preparations Dangerous for Supply* gives advice.

The safety data sheet is invaluable in enabling an employer to carry out a COSHH assessment. However, it is not a substitute for an assessment. Safety data sheets will describe the hazards of the chemicals, but only the user can assess the risk in the workplace.

Safety data sheets must be provided whether the chemical is sold in bulk or in packages. However, they do not have to be provided when chemicals are sold for private use through shops, due to the existence of comparable legislation covering such cases.

Labelling

Packaging containing dangerous chemicals must be properly labelled. The aim is to inform anyone handling the package or using the chemical about its hazards and to advise on the precautions. For workers the label is the supplement to information provided by the employer; for others it is a major way of propagating information. The label must always contain details about:

(a) the supplier;
(b) the chemical;
(c) the category of danger; and
(d) risk phrases and safety phrases.

Risk phrases and safety phrases are standard phrases set out in the CHIP Regulations. *Risk phrases* describe the dangers of the chemicals in more detail, eg 'May cause cancer' or 'Toxic by inhalation'. *Safety phrases* tell the user what to do or not to do with the chemical, eg 'Keep away from children' or 'Do not empty into drains'.

A *warning symbol* is also required on most labels, eg a skull and crossbones or a picture of an explosion. Suppliers are responsible for using the correct label.

Labelling follows classification. For substances, the first step is to look in the Approved Supply List. Where the substance is not listed, or if the chemical is a preparation, then the label is derived from the classification using the advice in the Approved Guide to the Classification and Labelling of Substances and Preparations Dangerous for Supply.

Packaging

The Regulations require chemicals to be packaged safely to withstand the conditions of supply.

HAZARDOUS SUBSTANCES AND PREPARATIONS – CLASSIFICATION

The following classifications are detailed in Schedule 1 of the CHIP Regulations.

Category of danger	Property (See Note 1)	Symbol – letter
1. Physico-Chemical Properties		
Explosive	Solid, liquid, pasty or gelatinous substances and preparations which may react exothermically without atmospheric oxygen, thereby quickly evolving gases. Under defined test conditions these can detonate, quickly deflagrate or explode if heated while being partially confined.	E
Oxidising	Substances and preparations which give rise to an exothermic reaction in contact with other substances, particularly those which are flammable.	0
Extremely flammable	Liquid substances and preparations having an extremely low flash point and a low boiling point. Also, gaseous substances and preparations which are flammable in contact with air at ambient temperature and pressure.	F+
Highly flammable	The following substances and preparations: (a) substances and preparations which may become hot and finally catch fire in contact with air at ambient temperature without any application of energy; (b) solid substances and preparations which may readily catch fire after brief contact with a source of ignition and which continue to burn or to be consumed after removal of the source of ignition; (c) liquid substances and preparations having a very low flash point; (d) substances and preparations which, in contact with water or damp air, evolve highly flammable gases in dangerous quantities. (See Note 2)	F
Flammable	Liquid substances and preparations having a low flash point.	None

2. Health Effects

Category	Description	Symbol
Very toxic	Substances and preparations which in *very low quantities* can cause death, acute or chronic damage to health when inhaled, swallowed or absorbed via the skin.	T+
Toxic	Substances and preparations which in *low quantities* can cause death, acute or chronic damage to health when inhaled, swallowed or absorbed via the skin.	T
Harmful	Substances and preparations which may cause death, acute or chronic damage to health when inhaled, swallowed or absorbed via the skin.	Xn
Corrosive	Substances and preparations which may, on contact with living tissues, *destroy* them.	C
Irritant	Non-corrosive substances and preparations which through immediate, prolonged or repeated contact with the skin or mucous membrane, may cause *inflammation*.	Xi
Sensitising	Substances and preparations which, if they are inhaled or penetrate the skin, are capable of eliciting a reaction by *hypersensitisation*, so that on further exposure to the substance or preparation, characteristic adverse effects are produced.	
Sensitising by inhalation		Xn
Sensitising by skin contact		Xi
Carcinogenic (see Note 3)	Substances and preparations which, if they are inhaled, ingested or penetrate the skin, may induce *cancer* or increase its incidence.	
Category 1		T
Category 2		T
Category 3		Xn
Mutagenic (See Note 3)	Substances and preparations which, if they are inhaled, ingested or penetrate the skin, may induce *hereditable genetic defects* or increase their incidence.	
Category 1		T
Category 2		T
Category 3		Xn
Toxic for reproduction (See Note 3)	Substances and preparations which, if they are inhaled, ingested or penetrate the skin, may produce or increase the incidence of *non-hereditable adverse effects* in the progeny, and/or an impairment of male or female reproductive functions or capacity.	
Category 1		T
Category 2		T
Category 3		Xn
Dangerous for the environment (See Note 4)	Substances which, if they were to enter into the environment, would/could present an immediate or delayed danger for one or more components of the environment.	N

Figure 10.1 Categories of danger

Notes
1. As further described in the *Approved Classification and Labelling Guide*.
2. Preparations packed in *aerosol dispensers* shall be classified as *flammable* in accordance with the additional criteria set out in Part II of this Schedule.
3. The categories are specified in the *Approved Classification and Labelling Guide*.
4. (a) In certain cases specified in the *approved supply list* and the *Approved Classification and Labelling Guide*, substances known to be *dangerous for the environment*, do not require to be labelled with the symbol for this category of danger.
   (b) This category of danger does not apply to preparations.

EU REGULATION CONCERNING THE REGISTRATION, EVALUATION, AUTHORISATION AND RESTRICTION OF CHEMICALS (REACH)

This European Regulation has the following objectives:

- to provide a high level of protection of human health and the environment from the use of chemical substances;
- to require those who place chemicals on the market, ie manufacturers and importers, to be responsible for managing and understanding the risks associated with their use;
- to allow the free movement of substances on the EU market;
- to enhance innovation in, and the competitiveness of, the EU chemicals industry; and
- to promote the use of alternative methods for the assessment of the hazardous properties of hazardous substances.

Under REACH, manufacturers and/or importers of substances are required to register such substances with a central European Chemicals Agency, a registration package being supported by a standard set of data on that substance. The regime applies to substances manufactured or brought into the EU in quantities of 1 tonne per year or more. As such, it applies to all individual chemical substances:

a. on their own;
b. in preparations; or
c. in articles, if the substance is intended to be released during normal and reasonably foreseeable conditions of use from an article.

PRINCIPLES OF TOXICOLOGY

Important definitions

Toxicology The study of the body's responses to toxic substances
Toxicity (1) The ability of a chemical molecule to produce injury once it reaches a susceptible site in or on the body (2) The quantitative study of the body's responses to toxic substances
Intoxication The general state of harm caused by the effects of a toxic substance
Detoxification The process in the body when decomposition of toxic substances occurs to produce harmless substances that are eliminated from the body.

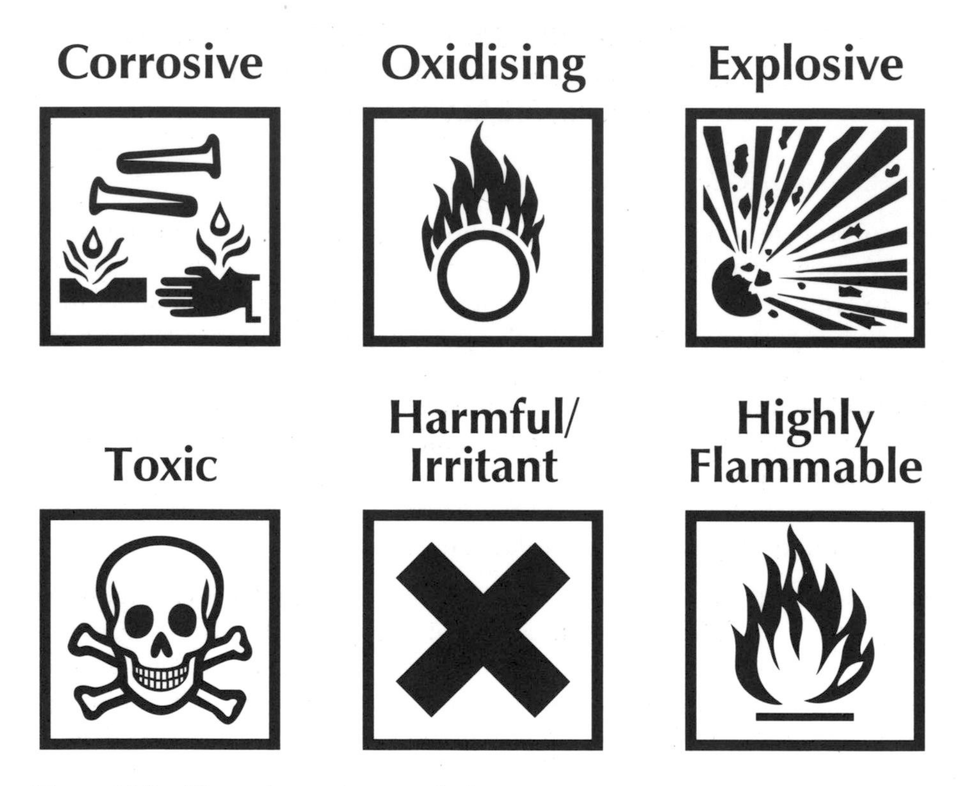

Figure 10.2 Hazard warning symbols

Effects of exposure to toxic substances

The effects on the body of exposure to toxic substances vary considerably, as outlined below.

1. Acute effect

A rapidly produced effect following a single exposure to an offending agent.

2. Chronic effect

An effect produced as a result of prolonged exposure or repeated exposures of long duration. Concentrations of the offending agent may be low in both cases.

3. Sub-acute effect

A reduced form of acute effect.

4. Progressive chronic effect

An effect that continues to develop after exposure ceases.

5. Local effect

An effect usually confined to the initial point of contact. The site may be the skin, mucous membranes of the eyes, nose or throat, liver, bladder, etc.

6. Systemic effects

Such effects occur in parts of the body other than at the initial point of contact, and are associated with a particular body system, eg the respiratory system, central nervous system.

Routes of entry of toxic substances into the body

1. Inhalation

Inhalation of toxic substances, in the form of a dust, gas, mist, fog, fume or vapour, accounts for approximately 90 per cent of all ill health associated with toxic substances. The results may be acute (immediate) as in the case of gassing accidents, eg chlorine, carbon monoxide; or chronic (prolonged, cumulative) as in the case of exposure to chlorinated hydrocarbons, lead compounds, benzene, numerous dusts (which produce pneumoconiosis), mists and fogs (such as that from paint spray, oil mist) and fume (such as that from welding operations).

2. Pervasion

The skin, if intact, is proof against most, but not all, inputs. Certain substances and micro-organisms are capable of passing straight through the intact skin into underlying tissue, or even into the bloodstream, without apparently causing any changes in the skin.

The resistance of the skin to external irritants varies with age, sex, race, colour and, to a certain extent, diet. Pervasion, as a route of entry, is normally associated with occupational dermatitis, the causes of which may be broadly divided into two groups:

(a) *primary irritants* – substances that will cause dermatitis at the site of contact if permitted to act for a sufficient length of time and in sufficient concentrations, eg strong alkalis, acids and solvents;

(b) *secondary cutaneous sensitisers* – substances that do not necessarily cause skin changes on first contact, but produce a specific sensitisation of the skin. If further contact occurs after an interval of, say, seven days or more, dermatitis will develop at the site of the second contact. Typical skin sensitisers are plants, rubber, nickel and many chemicals.

It should be noted that, for certain people, dermatitis may be a manifestation of psychological stress, having no relationship with exposure to toxic substances (an endogenous response).

3. Ingestion

Certain substances are carried into the gut, from which some will pass into the body by absorption. Like the lung, the gut behaves as a selective filter that keeps out many but not all harmful agents presented to it.

4. Injection/Implantation

A forceful breach of the skin, frequently as a cause of injury, can carry substances through the skin barrier.

Toxic substances are widely used in industry. Some indication of the most common hazards and the occupations associated with them are listed in the Social Security (Industrial Injuries) (Prescribed Diseases) Regulations 1985.

Dose–response relationship

A basic principle of occupational disease prevention rests upon the reality of threshold levels of exposure for the various hazardous agents below which humans can cope successfully without significant threat to their health. This concept derives from the quantitative characteristic of the dose–response relationship, according to which there is a systematic down change in the magnitude of a person's response as the dose of the offending agent is reduced.

Dose = Level of environmental contamination × Duration of exposure

With many dusts, for instance, the body's response is directly proportional to the dose received over a period of time: the greater the dose the more serious the condition, and vice versa. However, in the case of airborne contaminants, such as gases or mists, there is a concentration in air or dose below which most people can cope reasonably well. Once this concentration in air is reached (threshold dose) some form of body response will result. This concept is most important in the correct use and interpretation of Workplace Exposure Limits (formerly known as 'Threshold Limit Values').

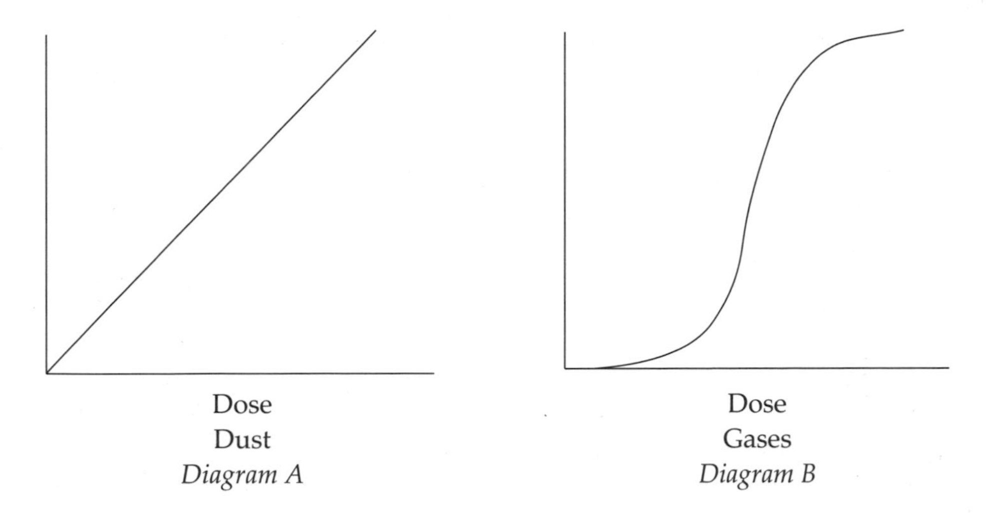

Diagram A

Diagram B

Diagram A shows a typical direct dose–response relationship, which is a feature of exposure to many dusts. In Diagram B the dose–response curve reaches a level of 'no response' at a point greater than zero on the dose axis. This cut-off point identifies the threshold dose that was the original basis for the setting of Threshold Limit Values.

Target organs and target systems

Certain substances have a direct or indirect effect on specific body organs (target organs) and body systems (target systems). Typical examples are:

Target organs Lungs, liver, brain, skin, bladder
Target systems Central nervous system, circulatory system, reproductive system.

Protective mechanisms – particulate matter

The protective mechanisms within the body respond largely according to the shape and size of particulate matter that may be inhaled. These mechanisms are as follows.

1. Nose

The coarse hairs in the nose, assisted by mucus from the nasal lining, act as a filter, trapping the larger particles of dust.

2. Ciliary escalator

The respiratory tract consists of the trachea (windpipe) and bronchi that branch to the lungs. The lining of the trachea consists of quite tall cells, each of which has a cilium growing from its head (ciliated epithelium). These cilia exhibit a wave-like motion so that a particle falling on to the cilia is returned back to the throat. Mucus assists these particles to stick.

3. Macrophages (Phagocytes)

Macrophages are wandering scavenger cells. They have an irregular outline and large nucleus, and can move freely through body tissue, engulfing bacteria and dust particles. They secrete hydrolytic enzymes that attack the foreign body.

4. Lymphatic system

This is the body's drainage system that acts as a clearance channel for the removal of foreign bodies, many of which are retained in the lymph nodes throughout the body. In certain cases a localised inflammation will be set up in the lymph node.

5. Tissue response

A typical example of tissue response is byssinosis ('Monday fever'), a chest condition of cotton workers, where the lung becomes sensitised to cotton dust through continuing exposure.

WORKPLACE EXPOSURE LIMITS

HSE Guidance Note EH40 *Workplace Exposure Limits* gives details of exposure limits that should be used for determining the adequacy of control of exposure of employees by inhalation to substances hazardous to health.

A Workplace Exposure Limit (WEL) is defined as the maximum concentration of an airborne substance, averaged over a reference period, to which employees may be exposed by inhalation. WELs are listed in HSE Guidance Note EH40.

The advice given in Guidance Note EH40 should be taken in the context of the requirements of the COSHH Regulations, especially:

- Regulation 6 (health risk assessment);
- Regulation 7 (control of exposure);
- Regulations 8 and 9 (use and maintenance of control measures);
- Regulation 10 (monitoring of exposure at the workplace).

Correctly applying the principles of good practice (see p 220) will mean exposures are controlled below the WEL. Additional guidance may be found in the COSHH General ACOP.

Units of measurement

The list of WELs given in the guidance note relates to personal exposure to substances hazardous to health in the workplace, unless stated otherwise.

Concentrations of gases and vapours in air are usually expressed as parts per million (ppm), a measure of concentration by volume, as well as in milligrams per cubic metre (mg/m^3) of air, a measure of concentration by mass.

Concentrations of airborne particles (fumes, dust, etc) are usually expressed in mg/m^3, with the exception of mineral fibres, which are expressed as fibres per millilitre of air.

HANDLING AND STORAGE OF DANGEROUS SUBSTANCES

A number of general principles can be considered in the safe handling and storage of toxic substances. Assuming that substitution as a control strategy has been considered and found impracticable then, in order of merit, the following practices and procedures should be followed:

(a) use in diluted form wherever possible;
(b) only limited quantities should be used or stored at any one time – large quantities should be stored in a purpose-built bulk chemical store;
(c) in certain cases containment of a specific area may be necessary – consider safe venting and drainage requirements;
(d) eliminate handling and dispensing from bulk – in certain processes the use of automatic systems, eg cleaning-in-place (CIP) systems, may be possible;
(e) provide adequate local exhaust ventilation (LEV) systems, which are subject to regular examination and test;
(f) in certain cases separation of substances may be necessary, eg acids from alkalis; with
(g) personal protective equipment (PPE) as an extra means, not sole means, of protection.

SAFETY DATA SHEET/PRODUCT INFORMATION

The CHIP Regulations specify that obligatory information under the following headings must be provided in a safety data sheet:

1. Identification of the substance/preparation.
2. Composition/information on ingredients.
3. Hazards identification.
4. First aid measures.
5. Fire fighting measures.

6. Accidental release measures.
7. Handling and storage.
8. Exposure controls / Personal protection.
9. Physical and chemical properties.
10. Stability and reactivity.
11. Toxicological information.
12. Ecological information.
13. Disposal considerations.
14. Transport information.
15. Regulatory information.
16. Other information.

CONTROL OF SUBSTANCES HAZARDOUS TO HEALTH (COSHH) REGULATIONS 2002

Industry uses some 40,000 different substances. Each one could be hazardous and will, therefore, fall within the scope of the COSHH Regulations. The regulations apply to nearly all companies and organisations, from the major chemical manufacturer to the smaller craft workshops, but the following industries are especially involved:

(a) major manufacturers and bulk users of chemical substances;
(b) users of substances in circumstances most likely to involve high exposure levels, eg spraying activities;
(c) the so-called 'dusty trades', including the ceramics and refractories industries, quarrying, foundries and metal manufacturing/ finishing processes; and
(d) users of processes which generate substances hazardous to health in appreciable quantities, eg certain welding, cutting, grinding, milling or sieving operations.

The principles outlined in the COSHH Regulations are supported by a number of ACOPs and Guidance Notes issued by the HSE.

Principal requirements of the COSHH Regulations

1. The employer must take account of the properties of substances used at work, in particular the EEC Inventory of over 5,000 substances and the EEC Directive on hazardous agents.
2. Employers have the following duties:
   (a) to make health risk assessments;
   (b) to control exposures;
   (c) to carry out monitoring; and

(d) to arrange for health surveillance, in particular:
  (i) health examinations;
  (ii) environmental monitoring; and
  (iii) the maintenance of control measures, eg exhaust ventilation systems.

Practical implementation of the regulations implies that the 'responsible person' – ie the employer or controller of premises, manufacturer and/or supplier of substances – must know about the potential for harm to employees or others of any substances used, sold, supplied or disposed of as part of his undertaking.

A 'substance' is defined as 'any natural or artificial substance, whether solid, liquid, gas or vapour, and includes human pathogens'. Lead and asbestos are specifically excluded from the regulations.

A 'substance hazardous to health' means any substance (including any preparation) which is:

(a) listed in Part I of the approved supply list as dangerous for supply within the meaning of the CHIP Regulations and classified as very toxic, toxic, harmful, corrosive or irritant;
(b) a substance specified in Schedule 1 (which lists substances assigned maximum exposure limits) or for which the HSC has approved an occupational exposure standard; (see HSE Guidance Note EH40 *Workplace Exposure Limits*);
(c) a biological agent;
(d) dust of any kind, when present in a substantial concentration in air; and
(e) any other substance arising from work which may be hazardous to health.

HEALTH RISK ASSESSMENTS

Perhaps the most significant duty on the employer under the regulations is the requirement to make a health risk assessment. Such an assessment should identify:

(a) the risks posed to the health of the workforce;
(b) the measures necessary to control exposure to those hazards; and
(c) other action that may be necessary to achieve compliance with Regulations 8–12.

Moreover, the assessment must be reviewed forthwith if there is reason to suspect that the assessment is no longer valid, or there has been a significant change in the work to which the assessment relates and where, as a result of the review, changes in the assessment are required, those changes must be made.

The starting point for most assessments is the hazard data information supplied by the manufacturer, supplier or importer, which must be provided in accordance with Section 6 of HASAWA. While there is no standard format for a health risk assessment, the following information should be available at the completion of the exercise:

(a) the risk involved in the use of the substance (toxic, corrosive, harmful or irritant), together with the route(s) of entry of that substance into the body, eg by inhalation of its vapour;
(b) the engineering or other controls necessary to prevent health risks arising;
(c) health and/or medical surveillance procedures necessary;
(d) environmental monitoring procedures necessary to be undertaken in the workplace;
(e) extent of the information, instruction, training and supervision necessary for operators using the substance; and
(f) the record keeping requirements, eg maintenance and test of local exhaust ventilation systems.

A specimen health risk assessment document is shown in Figure 10.3.

Company compliance procedures

Procedures should take place in two stages, thus:

1. Preliminary procedure

(a) Prepare an inventory of all the substances used for production, maintenance, cleaning and laboratory analysis.
(b) Identify the point of use for each material.
(c) Ensure information on each substance is adequate. If the information is inadequate, the manufacturer/supplier should be required to provide more comprehensive information.
(d) Once the list is established, a programme of assessment and control of hazardous substances should be commenced.

2. Final procedure

(a) Identify the risks through health risk assessment procedures.
(b) Assess and evaluate the risks.
(c) Control exposures.
(d) Ensure use of control measures.
(e) Maintain control measures.
(f) Institute workplace monitoring.
(g) Institute health surveillance procedures.
(h) Provide information, instruction and training for all staff on a regular basis, and at the induction stages for new employees.

HEALTH RISK ASSESSMENT
CONTROL OF SUBSTANCES HAZARDOUS TO HEALTH REGULATIONS

This Health Risk Assessment should be undertaken taking into account the supplier's safety data provided in accordance with the Chemicals (Hazard Information and Packaging for Supply) Regulations

---

Assessment No

Location	**Process/Activity/Use**

Substance Information

Name of substance	Chemical composition
Supplier	

Risk Information

Risk classification	Toxic / Corrosive / Harmful / Irritant Other classifications eg Flammable
Workplace Exposure Limits	WELS LTEL STEL
Route(s) of entry	Acute / Chronic / Local / Systemic effects
Exposure situations	
Exposure effects	
Estimate of potential exposure	Frequency of use
Quantities used	Duration of use

Storage Requirements
Small scale storage
Bulk storage

Air monitoring requirements and standards

First Aid Requirements

Health Surveillance Requirements

Routine Disposal Requirements

Procedure in the event of spillage

- **Small scale spillage**
- **Large scale spillage**

Information, Instruction and Training Requirements and Arrangements
Supervision Requirements

GENERAL CONCLUSIONS AS TO RISK

High/Medium/Low Risk
Prevention/Control of Exposure Requirements

____________________ **Assessor** **Date** __________________

Date of next review of assessment ____________________________

Figure 10.3 Specimen health risk assessment

It should be appreciated by employers that health records relating to individual exposures may have to be kept for 30 years.

Reference should be made to the COSHH action plan shown in Figure 10.4 when considering an approach to compliance with the COSHH Regulations.

PRINCIPLES OF GOOD PRACTICE

The COSHH (Amendment) Regulations 2004 specify the following principles of good practice with respect to the control of substances hazardous to health:

(a) design and operate processes and activities to minimise emission, release and spread of substances hazardous to health;
(b) take into account all relevant routes of exposure – inhalation, skin absorption and ingestion – when developing control measures;
(c) control exposure by measures that are proportionate to the health risk;
(d) choose the most effective and reliable control options that minimise the escape and spread of substances hazardous to health;
(e) where adequate control of exposure cannot be achieved by other means, provide, in combination with other control measures, suitable personal protective equipment;
(f) check and review regularly all elements of control measures for their continuing effectiveness;
(g) inform and train all employees on the hazards and risks from the substances with which they work and the use of control measures developed to minimise the risks;
(h) ensure that the introduction of control measures does not increase the overall risk to health and safety.

HEALTH SURVEILLANCE

Health surveillance is the regular review of the health of employees exposed to various forms of health risk, for instance from hazardous substances, manual handling operations or as a result of working on specific processes.

Legal requirements for health surveillance

While there is a general duty on an employer under section 2(1) of the HASAWA 'to ensure, so far as is reasonably practicable, the health, safety and welfare at work of all his employees', more specific duties to provide same can be found in regulations made under the Act, such as the COSHH Regulations and the MHSWR.

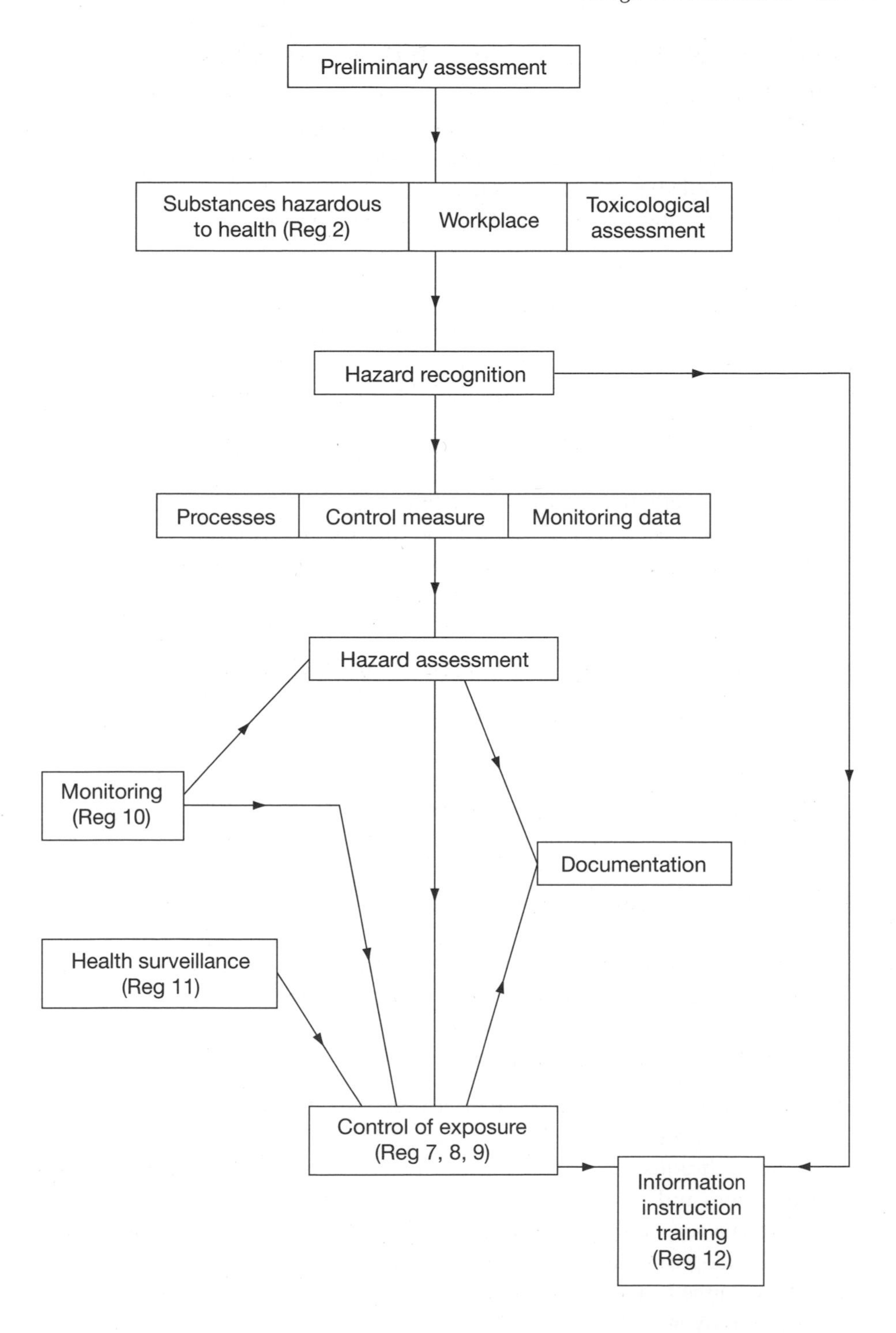

Figure 10.4 COSHH action plan

Control of Substances Hazardous to Health Regulations 2002

Under these regulations there is a broad duty on an employer to ensure the provision of suitable health surveillance where appropriate (Regulation 11). More specifically, health surveillance is necessary where an employee is exposed to a substance or process listed in Schedule 5 to the regulations, eg vinyl chloride monomer (VCM), potassium or sodium chromate or dichromate.

Secondly, where:

(a) there is an identifiable disease or adverse health effect that may be related to the exposure;
(b) there is a reasonable likelihood that this may occur under the particular work conditions; and
(c) there are valid techniques for detecting indications of it,

then health surveillance must be provided.

In this latter case a typical identifiable disease is dermatitis, occasionally encountered among operators in many industries through exposure to detergents, detergent sanitisers, solvents and oils.

Management of Health and Safety at Work Regulations 1999

Regulation 6 states quite categorically:

> 'Every employer shall ensure that his employees are provided with such health surveillance as is appropriate having regard to the risks to their health and safety which are identified by the [risk] assessment.'

The ACOP qualifies the above duty thus:

> 'The risk assessment will identify circumstances in which health surveillance is required by specific health and safety regulations, eg COSHH. Health surveillance should also be introduced where the assessment shows the following criteria to apply:

(a) there is an identifiable disease or adverse health condition related to the work concerned; and
(b) valid techniques are available to detect indications of the disease or condition; and
(c) there is a reasonable likelihood that the disease or condition may occur under the particular conditions of work; and
(d) surveillance is likely to further the protection of the health and safety of the employees to be covered.'

HSE guidance accompanying the regulations provides further information on the subject:

> 'Where appropriate, health surveillance may also involve one or more health surveillance procedures depending on suitability in the circumstances. Such procedures can include:

(a) inspection of readily detectable conditions by a responsible person acting within the limits of their training and experience;
(b) enquiries about symptoms, inspection and examination by a qualified person such as an occupational health nurse;
(c) medical surveillance, which may include clinical examination and measurement of physiological or psychological effects by an appropriately qualified person;
(d) biological effect monitoring, ie the measurement and assessment of early biological effects such as diminished lung function in exposed workers; and
(e) biological monitoring, ie the measurement and assessment of workplace agents or their metabolites either in tissues, secreta, excreta, expired air or any combination of these in exposed workers.

The primary benefit, and therefore objective of health surveillance should be to detect adverse health effects at an early stage, thereby enabling further harm to be prevented. The results of health surveillance can provide a means of:

(a) checking the effectiveness of control measures;
(b) providing feedback on the accuracy of the risk assessment; and
(c) identifying and protecting individuals at increased risk because of the nature of their work.'

Once it is decided that health surveillance is appropriate, such health surveillance should be maintained during the employee's employment unless the risk to which the worker is exposed and associated health effects are short term. The minimum requirement for health surveillance is the keeping of an individual health record.

Health surveillance procedures

Where it is appropriate, health surveillance may also involve one or more procedures depending upon their suitability in the circumstances. Such procedures can include:

(a) inspection of readily detectable conditions by a responsible person acting within the limits of their training and experience;
(b) enquiries about symptoms, inspection and examination by a qualified person such as an occupational health nurse;

(c) medical surveillance, which may include clinical examination and measurements of physiological or psychological effects by an appropriately qualified practitioner;
(d) biological effect monitoring, ie the measurements and assessment of early biological effects such as diminished lung function in exposed workers;
(e) biological monitoring, ie the measurement and assessment of workplace agents or their metabolites either in tissues, secreta, excreta, expired air or any combination of these in exposed workers.

Health surveillance is a very broad field and must be viewed as part of a general occupational health strategy aimed at protecting the health of people at work. Much will depend upon the risks to which people are exposed. What is important is that the form of health surveillance undertaken should be appropriate to these risks.

Health surveillance should be carried out by suitably qualified persons, such as a doctor with, preferably, a specialist qualification in occupational medicine (an occupational physician) or an occupational health nurse. It may involve the assessment of hazardous substances or their by-products in the body (by the examination of urine or blood) or of body functions (eg blood pressure, lung function). In some cases clinical examinations or tests may be necessary. Where medical examinations and inspections are called for employers must provide suitable facilities on site.

Under the COSHH Regulations, where health surveillance is required, records must be kept listing the personal details of the employee, eg name, sex, age and the jobs he has done and is doing that could involve exposure to hazardous substances. The records, or copies of them, must be kept for at least 40 years after the date of the last entry.

SUMMARY

1. A wide range of potentially dangerous substances is used in industry.
2. Reference to the descriptive phrases specified in the CHIP Regulations should be made when considering the relative hazards of substances.
3. The route of entry of a hazardous substance is significant in its potential for harm.
4. Procedures should be established for the handling and storage of dangerous substances, including systems for the acquisition of information, and the training of employees in hazards associated with the use of dangerous substances.

5. The COSHH Regulations place significant duties upon employers and users of dangerous substances, together with manufacturers and suppliers. Procedures for ensuring compliance with the COSHH Regulations should be established, commencing with an inventory of all chemicals currently in use, and the implementation of continuing monitoring and assessment procedures.
6. Employers need to understand the principles of toxicology in the assessment of health risks arising from hazardous substances.

CONCLUSION TO PART 3

The HASAWA has brought about great improvements in occupational safety. Improvements in the occupational health field have, however, been much slower, and the number of days lost through occupational disease and ill health remains unacceptably high, with substantial costs to companies and the nation as a whole.

There is a need, therefore, for organisations to examine their sickness absence rates, to control sickness absence more effectively, and to use the services of occupational health practitioners in order to reduce the toll of ill health associated with the work that people do.

Most companies use dangerous substances in one form or another, and the uncontrolled use of these substances can result in health risks to staff. The COSHH Regulations have the principal objective of eliminating or controlling these risks in order to protect the health of workers. The establishment and implementation of procedures to comply with these regulations is essential, including health risk assessments for all identified dangerous substances, health surveillance, environmental monitoring and the development of control measures. Where a company does not have trained professionals to undertake this work, they will have to buy in this expertise from outside. Training of operators in the safe use of dangerous substances is essential.

REFERENCES

Atherley, G R C (1978) *Occupational Health and Safety Concepts*, Applied Science Publishers, London

Chemicals (Hazard Information and Packaging for Supply) Regulations 2002, HMSO, London

Control of Substances Hazardous to Health Regulations 2002, The Stationery Office, London

Health and Safety Commission (1999) *Approved Code of Practice: Control of biological agents*, HMSO, London

Health and Safety Commission (1999) *Approved Code of Practice: Control of carcinogenic substances*, HMSO, London
Health and Safety Commission (1999) *Approved Code of Practice: Control of substances hazardous to health*, HMSO, London
Health and Safety Executive (1999) *COSHH Assessments*, HMSO, London
Health and Safety Executive (under constant revision) *Environmental Hygiene Guidance Note Series: EH/18 Toxic Substances: A precautionary policy; EH/22 Ventilation of the workplace; EH/26 Occupational Skin Diseases: Health and safety precautions; EH/40 Workplace Exposure Limits; EH/42 Monitoring Strategies for Toxic Substances; EH/44 Dust in the Workplace: General principles of protection*, HMSO, London
Sax, N I (1979) *Dangerous Properties of Industrial Materials*, Reinhold, New York
Stranks, J (2005) *Handbook of Health and Safety Practice*, Prentice Hall, London

Part 4

Safety Technology

11

Engineering Safety

The relative safety aspects of machinery, plant and equipment have always been one of the more significant areas of health and safety enforcement and practice. The law relating to machinery and equipment is covered by the Provision and Use of Work Equipment Regulations (PUWER) 1998. Only two terms are defined in these regulations:

Work equipment means any machinery, appliance, apparatus or tool, and any assembly of components which, in order to achieve a common end, are arranged and controlled so that they function as a whole.

Use in relation to work equipment means any activity involving that equipment, including starting, stopping, programming, setting, transporting, repairing, modifying, maintaining, servicing and cleaning.

Work equipment, thus, can include a dumper truck, a power press, a ladder, a hammer, an electric drill and a tractor. The definition is extremely broad. PUWER covers a number of general duties and states that every employer:

(a) shall ensure work equipment is suitable (Regulation 5);
(b) shall ensure that work equipment is maintained in an efficient state, in efficient working order and in good repair (Regulation 6);
(c) shall pay attention to any specific risks created by the work equipment (Regulation 7);
(d) shall provide users with adequate health and safety information and, where appropriate, written instructions pertaining to the use of the work equipment (Regulation 8);

(e) shall ensure that all users, managers and supervisors have received adequate health and safety training with particular reference to correct working methods, any risks involved and the precautions necessary in the use of the equipment (Regulation 9);
(f) shall ensure that specific measures are taken to prevent risks arising from certain identified dangerous parts of machinery (Regulation 11); and
(g) shall take measures to prevent the exposure of any person to a specified hazard, such as work equipment catching fire or overheating (Regulation 12).

SELECTION OF WORK EQUIPMENT

Regulation 4 PUWER places an absolute duty on employers to ensure that work equipment is so constructed or adapted as to be suitable for the purpose for which it is used or provided. In selecting work equipment every employer shall have regard to the working conditions and the risks to the health and safety of persons that exist in the premises or undertaking in which that work equipment is to be used and any additional risk imposed by the use of that work equipment.

In particular, when selecting work equipment, employers should take account of ergonomic (work-related) risks (ACOP). Ergonomic design takes account of the size and shape of the human body and should ensure that the design is compatible with human dimensions. Operating positions, working heights, reach distances, etc can be adapted to accommodate the intended operator. Operation of the equipment should not place undue strain on the user. Operators should not be expected to exert undue force or stretch or reach beyond their normal strength or physical reach limitations to carry out tasks. This is particularly important for highly repetitive work such as working on supermarket checkouts or high speed 'pick and place' operations (HSE Guidance).

PRINCIPLES OF MACHINERY SAFETY

In any assessment of machinery safety, two factors must be considered:

(a) The mechanical factors, in terms of design, operation and reliability of guards and safety devices; and
(b) the human factors, with regard to physical operation of the machine (the ergonomic aspects are significant here), systems of work, the potential for operator error, routine maintenance procedures and waste removal.

Strategies directed at eliminating the hazards associated with machinery should take account of the principal causes of injury. For instance, a person may be injured while working with machinery through:

(a) coming into contact with it, or being trapped between the machinery and any material in or at the machinery or any fixed structure;
(b) being struck by, or becoming entangled in or by, any material in motion in the machinery;
(c) being struck by parts of the machinery ejected from it; and
(d) being struck by material ejected from the machinery.

(*Source:* BS EN ISO 12100 *Safety of Machinery*)

Machinery safety design features should take into account the following forms of machinery hazard with the principal objective of eliminating such hazards. Assessments of existing machinery should also take these hazards into account.

Traps

Traps in machinery, if unguarded, can result in a wide range of major injury accidents, including hand, arm and finger amputations. Traps take three principal forms:

(a) *In-running nips:* This form of trap is created where a moving belt or chain meets a roller or toothed wheel respectively, and at the point where revolving gears / toothed wheels, rollers or drums meet. Typical examples of in-running nips are shown in Figure 11.1.
(b) *Reciprocating traps:* These are encountered in the vertical and horizontal motion of certain machines, such as presses.

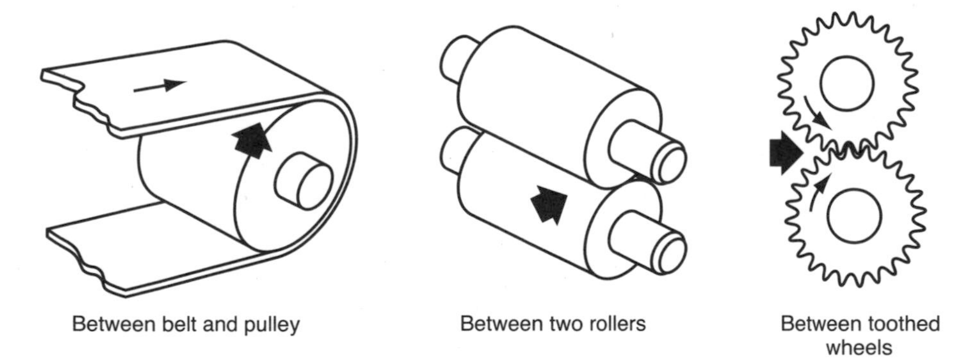

Figure 11.1 In-running nips
(Source: BS EN ISO 12100: 2002 'Safety of Machinery')

(c) *Shearing traps:* This form of machinery hazard produces a guillotine effect whereby a moving part traverses a fixed part of a machine or, alternatively, two moving parts traverse each other, with an action similar to that of garden shears. (See Figure 11.2.)

Entanglement hazards

The risk of the entanglement of clothing, hair and limbs can be present wherever unguarded rotating shafts, drills or chucks are in operation. (See Figure 11.3.)

Contact hazards

Contact with certain fixed and moving parts of machinery may cause injuries, eg with a hot surface, belt fastenings on a moving conveyor or the rough surfaces of a belt sander.

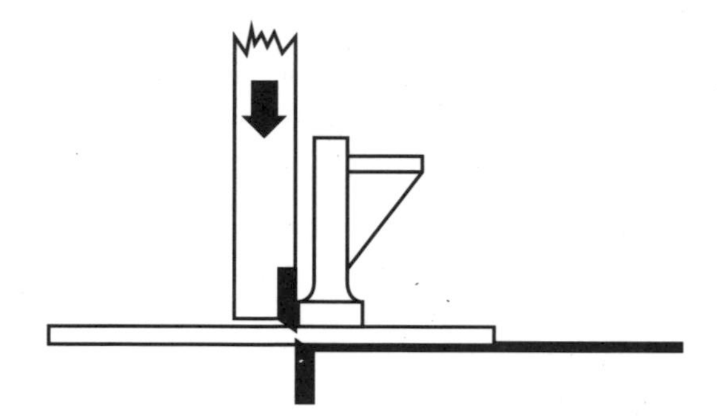

Figure 11.2 Shearing traps – a moving part traversing a fixed part
(Source: BS EN ISO 12100: 2002 'Safety of Machinery')

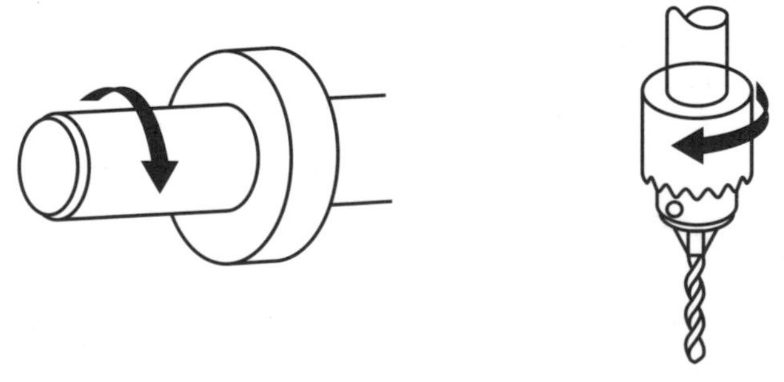

Figure 11.3 Entanglement risks
(Source: BS EN ISO 12100: 2002 'Safety of Machinery')

Ejection hazards

Some machines actually emit particles of metal. Grinding machines are a classic example, where there is a risk of ejection of metal particles, particularly into the face and eyes or, alternatively, of the abrasive wheel actually bursting without warning.

OTHER HAZARDS ASSOCIATED WITH MACHINERY OPERATION

While the above items are classed as the principal machinery hazards, there are many other hazards which need consideration at the design stage of a machinery-based work system, and during the operation of machinery. In certain cases people not actually involved in machinery operation may be exposed to risk of injury. Any assessment of the relative safety of a machine should take into account the following:

(a) procedures for job loading and removal;
(b) systems for changing of tools;
(c) safe removal of scrap and waste material;
(d) procedures for routine maintenance and adjustment, gauging, trying out following adjustment or setting, and in the event of breakdown;
(e) the potential for unexpected start-up or movement, uncovenanted stroke of the machine or mechanical failure;
(f) safe access to and egress from the machine and machine area;
(g) cleaning and housekeeping procedures in the machine area;
(h) availability of operating space; and
(i) potential risks to other persons passing through the machinery operating area.

MACHINERY GUARDS

BS EN ISO 12100 *Safety of Machinery* is the authoritative guidance on machinery safety in the United Kingdom, and is revised at regular intervals. The following forms of guard are specified in BS EN ISO 12100.

Fixed guard

This guard has no moving parts associated with it, or dependent upon the mechanism of any machinery, and, when in position, prevents access to a danger point or area. Generally, a fixed guard should not be removable other than through use of a hand tool. Fixed guards are generally used to guard transmission machinery, such as belt drives to machinery.

Adjustable guard

This is a guard which incorporates an adjustable element which, once adjusted, remains in that position during a particular operation. Adjustable guards are commonly used with band saws.

Distance guard

A distance guard does not completely enclose a danger point or area but places it out of normal reach. Tunnel guards to metal cutting machines are a typical example.

Interlocking guard

This is a guard which has a movable part so connected with the machinery controls that:

(a) the part(s) of the machinery causing danger cannot be set in motion until the guard is closed;
(b) the power is switched off and the motion braked before the guard can be opened sufficiently to allow access to the dangerous parts; and
(c) access to the danger point or area is denied while the danger exists.

Interlocking guard systems may take a number of forms, ie mechanical, electrical, hydraulic, pneumatic or a combination of these forms. Such guards should 'fail to safety'. Failure to safety (fail-safe) is defined in BS EN ISO 12100 as implying that any failure in, or interruption of, the power supply will result in the prompt stopping or, where appropriate, stopping and reversal of the movement of the dangerous parts before injury can occur, or the safeguard remaining in position to prevent access to the danger point or area.

Automatic guard

This guard is associated with, and dependent upon, the mechanism of the machinery and operates so as to remove physically from the danger area any part of a person exposed to the danger. This type of guard is commonly used with power presses. Certain automatic guards may be self-adjusting in that they prevent accidental access of a person to a danger point or area but allow the access of a workpiece which itself acts as part of the guard, the guard automatically returning to its closed position when the operation is completed. Portable circular saws frequently incorporate this form of self-adjusting automatic guard.

MACHINERY SAFETY DEVICES

A safety device is a protective appliance, other than a guard, which eliminates or reduces danger before access to a danger point or area can be achieved. Such devices take a number of forms.

Trip devices

This is a means whereby an approach by a person beyond the safe limit of working machinery causes the device to actuate and stop the machinery or reverse its motion, thus preventing or minimising injury at the danger point (BS EN ISO 12100). Trip devices can operate on a mechanical, photo-electric or ultrasonic basis, or through pressure-sensitive systems, such as pressure-sensitive mats, which stop a machine immediately a person approaches the danger area.

Two-hand control devices

This is a device which requires both hands to operate the machinery controls, thus affording a measure of protection from danger to the operator. Such devices are featured in clicking presses in the footwear industry.

Overrun devices

A device which, used in conjunction with a guard, is designed to prevent access to machinery parts which are moving by their own inertia after the power supply has been interrupted, so as to prevent danger. Overrun devices may take the form of rotation sensing devices, timing devices and certain forms of braking system.

Mechanical restraint device

A device which applies mechanical restraint to a dangerous piece of machinery which has been set in motion owing to failure of the machinery controls or other parts of the machinery, so as to prevent danger. Such devices are commonly used on pressure die-casting machines.

PLANNED MAINTENANCE

Regulation 6 of PUWER requires that all work equipment shall be maintained 'in an efficient state, in efficient working order and in good repair'. Compliance with this absolute duty implies the operation of planned maintenance programmes. 'Planned maintenance' is defined in BS 3811: 1974 as: 'maintenance organised and carried out with forethought, control and the use of records to a predetermined plan'.

A planned maintenance programme should incorporate records which identify each item of work equipment, generally by serial number, the maintenance procedure to be undertaken, the frequency of maintenance and the individual responsible for ensuring the procedure is completed satisfactorily.

LIFTING OPERATIONS AND LIFTING EQUIPMENT

Under the Lifting Operations and Lifting Equipment Regulations (LOLER) 1998, lifting equipment provided for use at work should be:

(a) strong and safe enough for its use;
(b) marked with its safe working load;
(c) installed and positioned to minimise risks;
(d) used safely;
(e) thoroughly examined and, where appropriate, inspected by a competent person on an ongoing basis.

All equipment used at work for lifting and lowering loads, including attachments and accessories, but with the exception of escalators, is covered by the regulations.

The regulations apply to all employers or self-employed persons who provide lifting equipment for use at work, as well as those who have control over the use of lifting equipment.

An ongoing requirement of the legislation is that:

(a) lifting operations must be planned, supervised and carried out in a safe manner by a competent person;
(b) equipment used for lifting people must be marked accordingly, and should be safe for such a purpose;
(c) before being used for the first time the equipment should, where appropriate, be thoroughly examined;
(d) it may also need to be thoroughly examined in use at set intervals, eg six months for accessories and equipment used for lifting people;
(e) all examinations must be carried out by a competent person;

(f) a report must be submitted by the competent person to the employer following a thorough examination or inspection of the equipment.

MACHINERY SAFETY ASSESSMENT

In assessing the relative safety of machines, the following form of audit should be undertaken:

1. Are machine controls correctly designed, safely located and clearly identified?
2. Is there an efficient stopping device for use in the event of an emergency? Is this device clearly identified?
3. Do the guards prevent access to danger points/areas? Do the safety devices operate effectively?
4. Do the features of the guarding system fail to safety? Are they 100 per cent reliable?
5. Can the safeguard be misused or defeated?
6. Does all electrical equipment comply with BS 2771?
7. Are there clearly identified access points for maintenance, lubrication and inspection?
8. What are the foreseeable machine failures? Does the guarding system cope effectively with such failures?
9. Is the machine correctly located so that other workers are not exposed to risk of injury caused by congestion of the working area?
10. Does the machine present secondary risks, such as noise or chemical hazards?
11. Does the machine create environmental pollution in the workplace, eg dust, fumes, gases? What controls are fitted to prevent such emissions?
12. Are temperature, lighting and ventilation levels in the machining area adequate and well maintained?

SUMMARY

1. There is a legal duty on the employer, both under the HASAWA and PUWER, to provide safe machinery and plant.

2. Assessment of machinery safety must take into account mechanical factors and human factors.

3. The principal machinery hazards are associated with traps, entanglement, contact and ejection risks.

4. Secondary hazards can be created by failure to take into account the total system for machine operation.
5. BS EN292 'Safeguarding of machinery' provides authoritative guidance on all aspects of machinery safety.
6. Machinery guards and safety devices should be properly maintained at all times.
7. New machinery and plant should be assessed for compliance with BS ISO 12100. Manufacturers, suppliers and importers of machinery and plant have clearly identified duties under Section 6 of the HASAWA.

REFERENCES

British Standards Institution (2002) *BS EN ISO 12100 Safety of Machinery*, BSI, Milton Keynes

Health and Safety Executive *Guidance Note Series on Plant and Machinery*, HMSO, London

Health and Safety Executive (1998) *Lifting Operations and Lifting Equipment Regulations 1998 and Guidance on Regulations*, HMSO, London

Health and Safety Executive (1998) *Provision and Use of Work Equipment Regulations 1992 and Guidance on Regulations*, HMSO, London

Stranks, J (1995) *Safety Technology*, Pitman, London

12

Fire Prevention

Fire is one of the natural elements, and generally taken to mean 'a state of burning or combustion'. Losses to industry through fire every year are substantial and, in order to prevent the deaths and human injury, damage to property and the consequent losses, it is essential that all employees at all levels in organisations are familiar with the causes of fire, fire protection procedures and the dangers associated with flammable substances.

THE FIRE TRIANGLE

For combustion to take place and continue, three basic requirements are necessary. These are:

(a) fuel, in solid, liquid or gaseous form;
(b) an ignition source; and
(c) air.

These three components of fire are shown in the Fire Triangle (see Figure 12.1).

If one of these three components is removed or not present, combustion cannot take place. Fire prevention strategies are, therefore, based on this fact, ie:

(a) limitation of, and control over, flammable substances stored on site;
(b) elimination of potential ignition sources; and
(c) restrictions on air available for combustion.

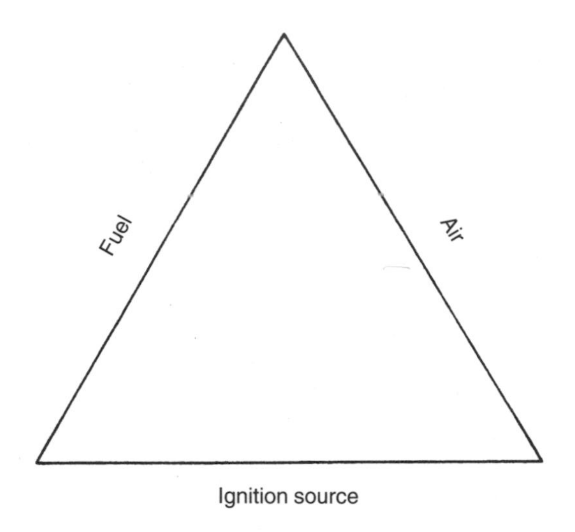

Figure 12.1 The fire triangle

CAUSES OF FIRE SPREAD

The two principal areas where fire can spread are production areas and storage areas. In production areas will be found electrical equipment, processing plant and equipment. Frictional heat and sparks from these items can set fire to packaging materials, waste materials, goods being processed and even dust in substantial quantities. In storage areas, which tend to be sparsely occupied, fire spread can be caused as a result of illicit smoking, through intruders – particularly children – committing acts of arson, or through defective electrical equipment setting fire to stored goods, packaging materials and certain combustible elements of buildings, such as wooden floors.

The principal causes of fire spreading in buildings are:

(a) lack of fire-separating walls between specific areas;
(b) poor housekeeping, resulting in combustible waste being stored in production and storage areas;
(c) oils and fats, which flow when in a burning state;
(d) the presence of vertical and horizontal features, such as trunking, ducting, conveyors, lift shafts and staircases, all of which readily permit fire spread from one part to another;
(e) the presence of dust at high level, which can explode in certain concentrations in air or, alternatively, burn rapidly once alight; and
(f) fires spreading from production to stored items in the same area.

FIRE EXTINCTION

The following are the three principles of fire extinction.

Cooling

Cooling is the most commonly used method of putting out fires, using water. The application of water to a fire results in much of this water being converted to steam, thereby preventing much of the heat present, which would have supported combustion, being returned to the combustible material. A point is reached where the temperature drops to such an extent that the continuous ignition of combustible materials ceases. Water in spray form is the most efficient means of application to burning materials, as opposed to water jets or applying buckets of water to a fire.

Smothering

Smothering a fire, by preventing more oxygen flowing to the fire or by applying an inert gas to the fire, brings about a reduction in the air available for combustion to the point where no further combustion will take place.

Starvation

Starvation involves reducing the amount of fuel available for combustion. This is undertaken by removing the fuel from the fire, isolating the fire from the fuel source, and by reducing the quantity or bulk of fuel present.

FIRE APPLIANCES

There are five principal types of hand-held fire appliance. Their use will depend upon the type of fire to be tackled. The two principal classes of fire and the appropriate fire appliances are shown in the table below.

Class of fire	Description	Appropriate extinguisher
A	Solid materials, usually organic, with glowing embers	Water, foam, dry powder, vapourising liquid, carbon dioxide
B	Liquids and liquefiable solids: (i) miscible with water, eg methanol, acetone (ii) immiscible with water	Water, foam (but must be stable on miscible solvents) carbon dioxide, dry powder Foam, dry powder, carbon dioxide, vapourising liquid

Table 12.1 Fire and fire appliance classification

Water appliances

There are two types of water appliance – stored pressure and gas cartridge. The stored pressure type contains carbon dioxide under pressure, water being expelled when the trigger is pulled. In the gas cartridge type, carbon dioxide is stored under pressure in a small cylinder. On breaking the seal with the plunger, the gas released expels the water through the nozzle.

Foam appliances

These are of the chemical foam type or stored pressure type. In both cases, when discharged and applied to burning material, they form a blanket of incombustible foam which prohibits further air for combustion. They are best used for small liquid spillage fires or for fires in small oil tanks whereby the foam can completely blanket the surface of the burning liquid. Foam appliances should always be completely discharged.

Carbon dioxide appliances

Carbon dioxide is suitable for application to both Class A and Class B fires, and fires involving electrical equipment. Liquid carbon dioxide is stored in a cylinder and, when discharged through a horn under its own pressure, quickly converts to carbon dioxide snow. The snow is subsequently converted to carbon dioxide gas by the fire. Carbon dioxide is not recommended for fires involving flammable liquids, unless the fire is very small.

Dry powder appliances

These appliances are of the stored pressure or gas cartridge type, and are extremely effective in dealing with flammable liquid fires.

Vapourising liquid appliances

These appliances incorporate a cylinder containing a liquid under pressure with dry carbon dioxide or nitrogen. On striking the knob, the seal is pierced thereby allowing the pressure to expel the liquid. A commonly used vapourising liquid is bromochlorodifluoromethane (BCF), which is particularly effective on fires involving electrical equipment and can be used quite satisfactorily for Class A and B fires, as well as small fires involving burning liquids.

COLOUR CODING OF FIRE APPLIANCES

All fire appliances must be coloured red and incorporate a colour coded label as shown below.

Extinguisher	*Colour code*
Water	Red
Foam	Cream
Carbon dioxide	Black
Dry chemical powder	Blue
Vapourising liquid	Green

Table 12.2 Colour coding of fire appliances

FIRE PROTECTION SYSTEMS

Such systems operate on the basis of detecting the presence of fire at a very early stage. They operate by three principal systems:

(a) heat sensing, where the actual temperature or rise in temperature is detected;
(b) smoke detection; and
(c) flame detection.

In all cases, these protection systems are linked with a form of fixed installation, such as a water sprinkler system or carbon dioxide system.

FIRE AND RESCUE AUTHORITIES

The Fire and Rescue Services Act 2004, which revoked the Fire Services Act 1947, created Fire and Rescue Authorities (FRAs). Under the Act, a FRA has the power 'to do anything which is calculated to facilitate, or is conducive or incidental to, the discharge of any of its functions'.

Functions of FRAs

FRAs have certain 'core' functions with respect to making provision for fire safety and fire-fighting and with dealing with road traffic accidents and emergencies. Other functions include providing directions relating to particular fires and emergencies and responding to other eventualities. They are enabled to run training courses and to charge for their services in certain cases. The FRA can further act as a negotiating body and make provision for pensions.

The FRA has a duty to secure water supply and to liaise with water undertakers to ensure supplies of water, particularly in an emergency.

Powers of authorised employees of a FRA

Enforcement powers of FRAs and their officers are incorporated in Part VI of the Act.

In an emergency, an employee authorised in writing by a FRA may do 'anything he reasonably believes to be necessary':

- for the purpose of extinguishing or preventing fire or protecting life or property;
- for the purpose of rescuing people or protecting them from serious harm in the event of a road traffic accident;
- for dealing with any other type of emergency;
- for the purpose of preventing or limiting damage to property resulting from the above.

In particular, an authorised employee may:

- enter a premises or place, if necessary by force, without the consent of the owner or occupier;
- move or break into a vehicle without the consent of the owner;
- close a highway;
- stop and regulate traffic; and
- restrict the access of persons to a premises or place.

A person commits an offence if without reasonable excuse he obstructs or interferes with an employee of a FRA taking authorised action.

Obtaining information and investigating fires (Section 45)

Authorised employees have extensive powers with respect to obtaining information and investigating fires. They may at any reasonable time enter premises:

(a) for the purpose of obtaining information; and
(b) if there has been a fire in the premises, for the purpose of investigating the causes of the fire and why it progressed as it did.

An authorised officer, that is, an employee of a FRA who is authorised in writing by the authority, may apply to a JP for a warrant to enter premises for the purpose of the above, but is unable to do so or considers that he is likely to be unable to do so, otherwise than by force. (Section 45)

Supplementary powers of authorised officers (Section 46)

If an authorised officer exercises a power of entry under section 45(1)(a) above, he may:

(a) take with him any other persons, and any equipment, that he considers necessary;
(b) require any person present on the premises to provide him with any facilities, information, documents or records, or other assistance, that he may reasonably request.

Authorised officers have other supplementary powers comparable with those of enforcement officers under the HASAWA.

LEGAL REQUIREMENTS

The Regulatory Reform (Fire Safety) Order 2005 replaced all the former legislation dealing with fire safety. The order is enforced by Fire and Rescue Authorities, except in specified cases.

The order distinguishes between *relevant persons* and *responsible persons* and places responsibility for fire safety in a workplace on the *responsible person*.

Important definitions

Relevant person means:

(a) any person (including the responsible person) who is or may be lawfully on the premises; and
(b) any person in the immediate vicinity of the premises who is at risk from a fire on the premises, but does not include a firefighter who is carrying out his duties in relation to a function of a fire authority.

Responsible person means:

(a) in relation to a workplace, the employer, if the workplace is to any extent under his control; and
(b) in relation to any premises not falling within paragraph (a):
    (i) the person who has control of the premises (as occupier is otherwise) in connection with the carrying on by him of a trade, business or other undertaking (whether for profit of not); or
    (ii) the owner, where the person in control of the premises does not have control in connection with the carrying on by that person of a trade, business or other undertaking.

Special technical and organisational measures include:

(a) technical means of supervision;
(b) connecting devices;
(c) control and protection systems;
(d) engineering controls and solutions;
(e) equipment;
(f) materials;
(g) protective systems; and
(h) warning and other communication systems.

General fire precautions:

1. In this order, *general fire precautions* in relation to premises means, subject to paragraph (2):

(a) measures to reduce the risk of fire on the premises and the risk of the spread of fire on the premises;
(b) measures in relation to the means of escape from premises;
(c) measures for securing that, at all material times, the means of escape can be safely and effectively used;
(d) measures in relation to the means for fighting fires on the premises;
(e) measures in relation to the means for detecting fire on the premises and giving warning in case of fire on the premises;
(f) measures in relation to the arrangements for action to be taken in the event of fire on the premises, including:
    (i) measures relating to the instruction and training of employees; and
    (ii) measures to mitigate the effects of fire.

2. The measures referred to in paragraph 1 do not include *special, technical and organisational measures* required to be taken or observed in any workplace in connection with the carrying on of any work process, where those precautions:

(a) are designed to prevent is reduce the likelihood of fire arising from such a work process is reduce its intensity; and
(b) are required to be taken is observed to ensure compliance with any requirement of the relevant statutory provisions within the meaning given by section 53(1) of the HASAWA.

Duties of responsible persons

A responsible person must:

- take such general fire precautions as will ensure, so far as is reasonably practicable, the safety of any of his employees;

- in relation to relevant persons who are not his employees, take such general fire precautions as may reasonably be required in the circumstances of the case to ensure that the premises are safe;
- make a suitable and sufficient assessment of the risks to which relevant persons are exposed for the purpose of identifying the general fire precautions he needs to take to comply with the requirements and prohibitions imposed upon him by or under this order;
- consider implications of the presence of dangerous substances in the risk assessment process;
- review the risk assessment if no longer valid or there has been a significant change in the matters to which it relates;
- record the significant findings of the risk assessment and details of any group being especially at risk;
- not commence a new work activity involving a dangerous substance unless a risk assessment has been made and measures required by the order have been implemented;
- make and give effect to arrangements for the effective planning, organisation, control, monitoring and review of preventive and protective measures;
- record the arrangements in specified cases;
- where a dangerous substance is present, eliminate or reduce risks so far as is reasonably practicable;
- replace a dangerous substance or the use of a dangerous substance with a substance or process which eliminates or reduces risks so far as is reasonably practicable;
- where not reasonably practicable to reduce above risks, apply *measures* to control the risk and mitigate the detrimental effects of fire;
- arrange safe handling, storage and transport of dangerous substances and wastes;
- ensure premises are equipped with appropriate fire-fighting equipment and with fire detectors and alarms and that non-automatic fire-fighting equipment is easily accessible, simple to use and indicated by signs;
- take measures for fire-fighting in the premises, nominate competent persons to implement these measures and arrange any necessary contact with external services;
- ensure routes to emergency exits and the exits themselves are kept clear at all times;
- comply with specific requirements dealing with emergency routes, exits and doors and the illumination of emergency routes and exits in respect of premises;
- establish and, where necessary, give effect to appropriate procedures for serious and imminent danger and for danger zones, including safety drills, nomination of competent persons to implement the procedures and restriction of access to areas on the grounds of safety;

WHEN THE FIRE ALARM SOUNDS

1. Close the windows, switch off electrical equipment and leave the room, closing the door behind you.
2. Walk quickly along the escape route to the open air.
3. Report to the fire warden at your assembly point.
4. Do not attempt to re-enter the building.

WHEN YOU FIND A FIRE

1. Raise the alarm by ... (If the telephone is to be used, the notice must include a reference to name and location.)
2. Leave the room, closing the door behind you.
3. Leave the building by the escape route.
4. Report to the fire warden at the assembly point.
5. Do not attempt to re-enter the building.

Figure 12.2 Fire instructions notice

- ensure additional emergency measures are taken in respect of dangerous substances;
- ensure relevant information is made available to emergency services and displayed at the premises;
- in the event of an accident, incident or emergency related to the presence of a dangerous substance, take immediate steps to mitigate the effects of fire, restore the situation to normal, and inform relevant persons;
- ensure only those persons essential for the carrying out of repairs and other necessary work are permitted in an affected area;
- ensure that the premises and any facilities, equipment and devices are subject to a suitable system of maintenance and are maintained in an efficient state, in efficient working order and in good repair;
- appoint one or more competent persons to assist him in undertaking the preventive and protective measures, ensuring adequate co-operation between competent persons;
- ensure that competent persons have sufficient time to fulfil their functions and the means at their disposal are adequate having regard to the size of the premises, the risks and the distribution of those risks;
- ensure competent persons not in his employment are informed of factors affecting the safety of any person and are provided with the same information as employees;
- provide employees with comprehensible and relevant information on the risks identified in the risk assessment, preventive and protective

measures, the identities of competent persons for the purposes of evacuation of premises and the notified risks arising in shared workplaces;

- before employing a child, provide the parent with comprehensible and relevant information on the risks to that child, the preventive and protective measures and the notified risks arising in shared workplaces;
- where a dangerous substance is on the premises, provide employees with the details of any such substance and the significant findings of the risk assessment;
- provide information to employers and the self-employed from outside undertakings with respect to the risks to those employees and the preventive and protective measures taken;
- provide non-employees working in his undertaking with appropriate instructions and comprehensible and relevant information regarding any risks to those persons;
- ensure the employer of any employees from an outside undertaking working in or on the premises is provided with sufficient information with respect to evacuation procedures and the competent persons nominated to undertake evacuation procedures;
- ensure employees are provided with adequate safety training at the time of first employment, and on being exposed to new or increased risks arising from transfer or change of responsibilities, introduction of, or change in, work equipment, the introduction of new technology and the introduction of a new system of work or a change respecting an existing system of work; and
- in the case of shared workplaces, co-operate with other responsible person(s), take all reasonable steps to co-ordinate the measures he takes to comply with this order with the measures taken by other responsible persons, and take all reasonable steps to inform other responsible persons.

Duties of employees

An employee must:

- take reasonable care for the safety of himself and others who may be affected by his acts or omissions while at work;
- co-operate with his employer to enable him to comply with any duty or requirement imposed by this order; and
- inform his employer or any other employee with the specific responsibility for the safety of his fellow employees of any work situation which represents a serious and immediate danger to safety, and any other matter which represents a shortcoming in the employer's protection arrangements for safety.

Enforcement arrangements

Inspectors appointed under the order have powers similar to those under the HASAWA. They are empowered to serve three types of notice:

1. Alterations Notice
   Where premises constitute a serious risk to relevant persons or may constitute such a risk if any change is made to them or the use to which they are put, the enforcing authority may serve on the responsible person an Alterations Notice.
2. Enforcement Notice
   If the enforcing authority is of the opinion that the responsible person has failed to comply with any provision of this order or of any regulations made under it, the enforcing authority may serve on that person an Enforcement Notice.
3. Prohibition Notice
   If the enforcing authority is of the opinion that use of premises involves or will involve a risk to relevant persons so serious that use of the premises ought to be prohibited or restricted, the authority may serve on the responsible person a Prohibition Notice, such a notice to include anything affecting the escape of relevant persons from the premises.

SUMMARY

1. Fire is the principal cause of many deaths and substantial losses in British industry.
2. Everyone should be aware of the causes of fire and the ways that fire spreads in buildings.
3. The principal effects of fire appliances are cooling, smothering and starvation.
4. Staff should receive regular instruction in the selection and correct use of fire appliances, although principal emphasis must always be on evacuating a building in the event of fire. Everyone should know the colour coding of fire appliances.
5. Means of escape in the event of fire should be clearly established and indicated.
6. Fire instruction notices must be displayed at strategic points in workplaces.
7. Fire drills must be undertaken at least annually, and fire alarms sounded weekly.

8. Fire safety legislation in workplaces is covered by the Regulatory Reform (Fire Safety) Order 2005, which places a range of duties on defined 'responsible persons' with respect to 'relevant persons' including that of undertaking fire risk assessments.

9. Under the Order inspectors have powers to serve on the responsible person an Alterations Notice, an Improvement Notice and a Prohibition Notice.

REFERENCES

Regulatory Reform (Fire Safety) Order 2005, HMSO, London

13

Electrical Safety

PRINCIPLES OF ELECTRICAL SAFETY

The two principal hazards associated with the use of electricity are the risk of electrocution and shock. Fatal accidents associated with the use or misuse of electricity average 50 per annum with approximately 1,000 people being injured at work through shock and burns.

Legal requirements

The principal legal requirements relating to the safe use of electricity are the Electricity at Work Regulations 1989. Further guidance is given in the Memorandum of Guidance accompanying the regulations, a number of British Standards and the Regulations for Electrical Installations published by the Institution of Electrical Engineers (the IEE 'Wiring Regulations').

These regulations are made under the HASAWA. They impose duties on employers, the self-employed, eg electrical contractors, and employees ('duty holders') all of whom have equal levels of duty. The regulations apply to all places of work and cover all electrical equipment ranging from battery-operated equipment to high voltage installations. They are framed in fairly general terms, which lay down principles of electrical safety, as opposed to the former regulations which went into great detail. Broadly, they require precautions to be taken against the risk of death or personal injury from electricity in work activities, and cover systems, work activities and protective equipment, the strength and capability of electrical equipment, work in adverse or hazardous environments and the basic principles – insulation, protection and placing of conductors, earthing or other suitable precautions, connections, means for protecting from excess current, for cutting off the supply and for isolation, work on equipment made dead, work on or near live conductors, and the provision of working space, access and lighting.

Regulation 16 requires persons to be 'competent to prevent danger and injury', implying the need for the appointment by management of competent persons (see Chapter 4) to cover electrical work activities. The defence of 'all reasonable precautions and all due diligence' (see Chapter 1) is available to a person charged with an offence under the regulations.

Electrical safety, therefore, is primarily concerned with protecting people from electric shock, which could have fatal results, and from fire and burns arising from contact with electricity. Principal protective measures against electric shock are:

(a) protection against direct contact, ie by providing proper insulation for those parts of equipment liable to be electrically charged; and
(b) protection against indirect contact, eg by the provision of effective earthing for metallic enclosures which are liable to be charged with electricity if the basic insulation fails for any reason.

The following aspects are of significance in providing protection against electrocution and electric shock.

Earthing

Earthing implies connection to the general mass of earth in such a manner as will ensure at all times an immediate discharge of electrical energy without danger. Earthing, to give protection against indirect contact with electricity, can be achieved in a number of ways, including the connection of extraneous conductive parts of premises (radiators, taps, water pipes) to the main earthing terminal of the electrical installation. This creates an equipotential zone and eliminates the risk of shock that could occur if a person touched two different parts of the metalwork liable to be charged, under earth fault conditions, at different voltages. Where an earth fault exists, such as when a live part touches an enclosed conductive part, eg metalwork, it is vital to ensure that the electrical supply is automatically disconnected.

This disconnection is brought about by the use of overcurrent devices, ie correctly rated fuses or circuit breakers, or by correctly rated and placed residual current devices. The maintenance of earth continuity is crucial.

Fuses

A fuse is basically a strip of metal of such size as would melt at a predetermined value of current flow. It is placed in the electrical circuit and, on melting, cuts off the current to that circuit. Fuses should be of a type and rating appropriate to the circuit and the appliance it protects.

Circuit breakers

This device incorporates a mechanism that trips a switch from the 'ON' to 'OFF' position if an excess current flows in the circuit. A circuit breaker should be of the type and rating for the circuit and appliance it protects.

Earth leakage circuit breakers (residual current circuit breakers)

Fuses and circuit breakers do not necessarily provide total protection against electric shock. Earth leakage circuit breakers (ELCBs) provide protection against earth leakage faults, particularly at those locations where effective earthing cannot necessarily be achieved.

Reduced voltage

Reduced voltage systems are another form of protection against electric shock, the most commonly used being the 110 volt centre point earthed system. In this system the secondary winding of the transformer providing the 110 volt supply is centre tapped to earth, thereby ensuring that at no part of the 110 volt circuit can the voltage to earth exceed 55 volts.

Safe systems of work

Where work is to be undertaken on electrical apparatus or a part of a circuit, a formally operated safe system of work should be used. This normally takes the form of a permit to work system which ensures the following procedures:

(a) switching out and locking off of the electricity supply, ie isolation;
(b) checking by use of an appropriate voltage detection instrument that the circuit or part of same to be worked on is dead before work commences;
(c) high levels of supervision and control to ensure the work is carried out correctly;
(d) physical precautions, such as the erection of barriers to restrict access to the area, are implemented; and
(e) formal cancellation of the permit to work once the work is completed and return to service of the plant or system in question.

CAUSES AND EFFECTS OF SHOCK

Electric shock results from an electric current flowing through the body. This has a direct effect on body organs and the central nervous system. The effect can be fatal if the heart rhythm is disturbed for long enough to stop blood flowing to the brain. Emergency action is vital in such cases. Resuscitation, ie mouth-to-mouth resuscitation, must be commenced quickly and be maintained until the patient recovers or, alternatively, is pronounced dead by a registered medical practitioner. The extent of the shock received is partly dependent upon the voltage of the current. However, the amperage is more significant than voltage, and an alternating current is more dangerous than a direct one. A current received through dry clothing is less dangerous than one received through wet clothing or directly on the bare skin and, obviously, all currents are more dangerous when the body is earthed than when insulation is provided by rubber-soled shoes or a rubber mat.

While the effects of electric shock can be fatal in many cases, non-fatal effects can include fractured bones and damage to flesh. For instance, the flesh at the point of entry may be damaged to a degree which ranges from a mild burn to severe destruction of muscles and internal organs. These injuries take a long time to heal because the mass of dead tissue takes a certain amount of time to separate itself from the living tissue. Sometimes the muscular spasm makes it impossible for the individual to let go of the object producing the shock and it passes for a longer period, producing damage to the brain or paralysis of the heart or respiration system. Since a state of suspended animation may last for some time, it is important that resuscitation be carried on, in many cases, for hours following receipt of the electric shock.

PORTABLE ELECTRICAL APPLIANCES

Deaths and injuries are commonly associated with portable electrical tools and equipment, such as welding equipment, battery chargers, drills, saws, and grinders and various domestic appliances, such as kettles, heaters and electric cookers, commonly used in workplaces. Hazards vary from the risk of electrocution and death to burns, shocks, eye injuries (from arc welding) and the ever-constant risk of explosion and fire due to the presence, as in battery charging, of hydrogen gas. Fires can result through the emission of sparks, arcing, short circuits, overloading of circuits or the breakdown of insulation on old wiring resulting in short circuiting.

Precautions

The precautions necessary in the use of portable electrical equipment can be related principally to the risk of injury to people and that of fire.

1. Flexible leads should be protected from mechanical damage.
2. The outer covering of a flexible lead should be firmly clamped at its end terminations to relieve strain on the inner conductors.
3. Apparatus should never be pulled or suspended from its lead.
4. The inner conductors of a flexible lead should always be properly connected into the appliance or into a plug of approved type of connector.
5. Any exposed metalwork on a portable appliance should normally be firmly connected to earth. A three-core flexible lead is essential. (Where the apparatus is of the double-insulated or all-insulated type an earthing terminal is not necessary.)
6. As far as possible reduced voltage portable appliances should be used, ie 110/55 volts or, in the case of hand lamps, 25/12.5 volts. The reduced voltage will reduce the severity of an electric shock.
7. Portable hand tools should be inspected and tested regularly by a competent person, using a standard test set.

HSE Guidance Note *Maintaining Portable and Transportable Electrical Equipment* (HS(G)107) provides further guidance on this matter.

Testing of portable electrical appliances

A principle of electrical safety is that there should be two levels of protection for the operator or user. This results in two classes of appliance. **Class 1** appliances incorporate both earthing and insulation (earthed appliances), whereas **Class 2** appliances are doubly insulated. The testing procedures for Class 1 and Class 2 appliances differ according to the type of protection provided.

Testing should be undertaken on a regular basis and should incorporate the following:

- inspection for any visible signs of damage to or deterioration of the casing, plug terminals and cable sheath;
- an earth continuity test with a substantial current capable of revealing a partially severed conductor;
- high voltage insulation tests.

The test results should be recorded, thus enabling future comparisons to determine any deterioration or degradation of the appliance. The control system should include:

- clear identification of the specific responsibility for appliance testing;
- maintenance of a log listing portable appliances, date of test and a record of test results;

- a procedure for labelling appliances when tested with the date for the next inspection and test.

Any appliance that fails the above tests should be removed from use. An estimation of the frequency of testing must take into account the type of equipment, its usage in terms of frequency of use and risk of damage, and any recommendations made by the manufacturer/ supplier.

SUMMARY

1. The risk of electrocution and electric shock are the main hazards associated with electricity, although, in lesser cases, there can be severe body burns, muscular damage and fractured bones.
2. Electricity, when improperly used, can also be a cause of fires.
3. Protection strategies are directed at protection against direct contact, through the provision of proper insulation, and the avoidance of indirect contact, by good standards of earthing.
4. Other important strategies for protecting people against electric shock include the use of correctly rated fuses, circuit breakers, reduced voltage and the implementation of safe systems of work, in particular permit to work systems for work on electrical systems.
5. The effects of electric current flowing through the body should be understood by employees, together with procedures for resuscitation.

REFERENCES

Beckingsale, A A (1976) *The Safe Use of Electricity*, RoSPA, Birmingham

Electricity at Work Regulations 1989 (SI 1989 No 635), HMSO, London

Health and Safety Executive (1980) *Electrical Testing: Safety in electrical testing* (HSE Booklet HS(G)13), HMSO, London

Health and Safety Executive (1989) *Memorandum of Guidance on the Electricity at Work Regulations*, 1989, HMSO, London

Health and Safety Executive (1991) *Guidance for Small Businesses on Electricity at Work*, HSE Information Centre, Sheffield

Health and Safety Executive (2004) *Maintaining Portable and Transportable Electrical Equipment* (HS(G)107), HSE Books, Sudbury

Health and Safety Executive Electricity Regulations (Booklet SHW 928), HMSO, London

Hughes, E (1978) *Electrical Technology*, Longman, London

Imperial College of Science and Technology (1976) *Safety Precautions in the Use of Electrical Equipment*, ICST, London

14

Structural Safety

A substantial number of accidents in the workplace are associated, both directly and indirectly, with structural features – ie floors, stairs, floor openings, entrance and exit points. Poor standards of structural safety can result in slips and falls, some of which can result in major injuries like fractured arms and legs, and even death in certain cases. Good design of the structural aspects of working areas together with regular preventive maintenance are, therefore, crucial to providing a safe workplace.

The Workplace (Health, Safety and Welfare) Regulations 1992 (the 'Workplace' Regulations), and the supporting ACOP and HSE Guidance, lay down the principal requirements relating to structural safety. As with the MHSWR and PUWER, the majority of these duties on employers under the Workplace Regulations are of an absolute nature.

FLOORS AND TRAFFIC ROUTES

Floors and traffic routes must be of sound construction. They must not be sloped, uneven or slippery. For cleaning and other tasks using water, floors must be effectively drained. So far as is reasonable, every floor and traffic route must be kept free from obstruction and from articles or substances which could cause people to slip, trip or fall. Sufficient handrails and/or guards must be provided on staircases (Regulation 12).

FALLS OR FALLING OBJECTS

Regulation 13 requires that suitable and effective measures be taken to prevent people from:

(a) falling a distance likely to cause injury; and
(b) being struck by a falling object likely to cause personal injury.

Tanks, pits or structures must be securely covered or fenced to avoid the risk of people falling into a dangerous substance.

WINDOWS, AND TRANSPARENT OR TRANSLUCENT DOORS, GATES AND WALLS

These must be constructed of safety material. They must also be appropriately marked, or incorporate design features to make them easily apparent (Regulation 14).

WINDOWS, SKYLIGHTS AND VENTILATORS

No operable window, skylight or ventilator should be opened, closed or adjusted in a manner which causes any person to risk his health or safety. Similarly, when opened, they must not be in a position which is likely to expose a person to risk, eg of falling out of the window (Regulation 15).

All windows and skylights must be designed, constructed and positioned so that they may be cleaned safely (Regulation 16).

ORGANISATION OF TRAFFIC ROUTES

There is a general duty to ensure the workplace is organised in such a way that pedestrians and vehicles can circulate in a safe manner. Traffic routes must be suitable – in size, number and position – for the persons or vehicles using them. All traffic routes must be well indicated (Regulation 17).

DOORS AND GATES

Doors and gates must be well constructed (including being fitted with any necessary safety devices) (Regulation 18).

ESCALATORS AND MOVING WALKWAYS

Escalators and moving walkways must:

(a) function safely;
(b) be equipped with any necessary safety devices; and
(c) be fitted with one or more emergency stop controls which are easily identifiable and readily accessible (Regulation 19).

ACCESS TO AND EGRESS FROM THE WORKPLACE

Section 2 of the HASAWA requires the occupier of the workplace to provide for his employees 'means of access to and egress from the workplace that are, so far as is reasonably practicable, safe and without risks to health'. This requirement applies to all employment situations, be the workplace an office block, factory, offshore platform or a coal face several miles below ground.

Typical risks associated with failure to comply with this requirement include defective and dangerous floors, staircases, catwalks, ladders, scaffolds and roadways, all of which present some form of risk to employees, visitors and members of the public who may use a premises such as a supermarket or public building.

WORK AT HEIGHT

The Work at Height Regulations 2005 impose health and safety requirements with respect to work at height, with certain exceptions including by instructors or leaders in recreational climbing and caving. The regulations apply to all work at height where there is a risk of a fall liable to cause personal injury

Work at height means:

(a) work in any place, including a place at or below ground level;
(b) obtaining access to or egress from such place while at work, except by a staircase in a permanent workplace,

where, if measures required by these regulations are not taken, a person could fall a distance liable to cause personal injury.

The regulations set out a simple hierarchy for managing and selecting equipment for work at height:

(a) the avoidance of work at height where possible;
(b) the use of work equipment or other measures to prevent falls where work at height cannot be avoided; and

(c) where the risk of a fall cannot be eliminated, the use of work equipment or other measures to minimise the distance and consequences of a fall should one occur.

Duty holders

The regulations place duties on employers, the self-employed and any person who controls the work of others, such as building owners, engineering managers who may contract others to work at height, to the extent they control the work.

Duty holders are responsible for:

- organising and planning of work at height;
- ensuring that persons at work are competent, or supervised by competent persons;
- steps to be taken to avoid risk from work at height, including taking account of a risk assessment under the Management of Health and Safety At Work Regulations;
- selection of equipment for work at height;
- ensuring provision of particular work equipment for work at height, such as guard rails, toe-boards, barriers, working platforms, safety nets, airbags or similar collective means of protection, and personal fall protection systems;
- avoidance of risks from fragile surfaces, falling objects and danger areas; and
- inspection of work equipment used for work at height and of places of work at height.

A duty holder must ensure that work is postponed while weather conditions endanger health or safety.

Fundamentally, a duty holder must do all that is reasonably practicable to prevent anyone falling.

Persons at work

Employees and any person working under the control of another must:

(a) report any activity or defect that could endanger the safety of himself or another person to the duty holder; and
(b) use the equipment supplied (including safety devices) properly, following any training and instructions.

Schedules

Schedules to the regulations lay down requirements for:

- existing places of work and means of access to or egress at height;
- guard rails, toe-boards, barriers and similar collective means of protection;
- working platforms;
- collective safeguards for arresting falls;
- all personal fall protection systems;
- ladders; and
- particulars to be included in a report of inspection.

TRAFFIC SYSTEMS

Poorly designed traffic systems can represent a serious hazard to employees and other persons visiting a premises. Well-controlled traffic systems around a workplace are, therefore, extremely important.

Features of traffic systems that are aimed at reducing the risk of personal injury, property damage and vehicular accidents, include the following:

(a) segregation of vehicular traffic routes from pedestrian routes;
(b) installation and operation of pedestrian barriers, pedestrian crossings and traffic lights at dangerous junctions;
(c) clear identification of vehicle parking areas, separating commercial vehicles from private cars, with directional signs at strategic locations;
(d) effective speed control over all vehicles, such as a 10 mph speed limit that is rigidly enforced and supported by 'sleeping policemen' or 'speed humps' which force traffic to slow down;
(e) operation of one-way systems where possible;
(f) control over unauthorised parking, parking in non-parking areas and generally 'unsafe parking' activities;
(g) the use of convex mirrors, located at strategic points, particularly where fork-lift traffic may cross recognised vehicle routes in or around the premises; and
(h) the provision and maintenance of high levels of artificial lighting at external access and egress points, loading bays, pedestrian walkways and parking areas. (It should be noted that good standards of artificial lighting can contribute greatly to improving site security.)

SUMMARY

1. Good standards of structural safety are essential for the prevention of accidents associated with slips, trips and falls.
2. Special attention must be paid to the use of underground rooms used other than for storage purposes.
3. The provision of safe access to and egress from the workplace is a legal requirement.
4. A well-organised traffic system is essential for the prevention of accidents involving people and vehicles.
5. External lighting is an important aid to the maintenance of good traffic systems and in the prevention of breaches of site security, including theft from the premises and from vehicles parked on the premises.

REFERENCES

Engineering Equipment Users Association (1973) *Factory Stairways, Ladders and Handrails*, EEUA Handbook No 7, EEUA, London

Health and Safety Commission (1992) *Workplace (Health, Safety and Welfare) Regulations 1992 and Approved Code of Practice*, HMSO, London

Health and Safety Commission (2005): *Work at Height Regulations 2005*, HMSO, London

Health and Safety Executive (1978) *Road Transport in Factories*, HMSO, London

Health and Safety Executive (1982) *Transport Kills*, HMSO, London

15

Construction and Contractors

The construction industry covers a wide field of operations, from the very large civil engineering projects such as motorway construction to, at the other end of the spectrum, self-employed tradesmen carrying out minor improvements and repairs to buildings. The extensive employment of casual labour, the relationship between the occupier of premises, contractors and subcontractors, and the very real problems that can arise when contractors are working on an existing site, contribute to the high incidence of accidents in this industry. It is essential, therefore, for companies to establish clearly written procedures to regulate the activities of contractors when carrying out work on their behalf.

CONSTRUCTION HAZARDS

The principal hazards in construction operations are associated with:

- falls from ladders, working platforms, pitched roofs;
- falls through fragile roofs, openings in flat roofs and floors;
- falls of materials;
- collapses of excavations;
- site transport activities;
- machinery and powered hand tools;
- housekeeping failures;
- fire;
- personal protective equipment.

CONTRACTORS AND THE LAW

Contracting activities are covered under both civil law and criminal law. Under civil law, ie the law giving rise to claims for damages, contractors employed by a company are at risk of civil liability in damages from their own employees and other people, including company employees, if they fail to take adequate care for their safety during their operations on company premises and such persons are injured as a result.

Under criminal law, eg HASAWA, the prime responsibility rests with the occupier of the premises or site concerned. The company, therefore, is subject to prosecution if there are breaches of statutory provisions on its premises if these breaches are caused by the contractor or one of his subcontractors. On this basis a company must take all reasonably practicable measures to ensure compliance with these provisions throughout its premises, ie it must make such arrangements which, if implemented, would ensure legal compliance, and then ensure that they are carried out. Moreover, if the making of such arrangements by the company is not reasonably practicable, due to the nature of the contractor's activities, the company must do its best to ensure that the contractor makes appropriate arrangements and then carries them out.

The legal provisions relating to safety, health and welfare in construction activities are outlined in:

- Construction (Head Protection) Regulations 1989;
- Construction (Design and Management) (CDM) Regulations 2007.

The general provisions of the HASAWA apply to all construction-related activities. In particular, contractors have obligations under Section 3 (duties of employers to other than their employees), and may have obligations under Section 4 not to endanger persons who are not their employees. In particular, duties under the CDM Regulations must be read in conjunction with the duties under the MHSWR.

Head protection

The Construction (Head Protection) Regulations 1989 impose duties on employers, people in control of construction sites, the self-employed and employees regarding the provision and use of head protection wherever there is a reasonably foreseeable risk of head injury. Employers must provide and maintain head protection, and ensure it is worn wherever there is risk of head injury. Persons in control of construction sites, eg main contractors, site managers, in conjunction with employers must identify when and where head protection must be worn, inform operators accordingly, and supervise the wearing of same.

Employees, similarly, have a duty to wear head protection in designated areas and operations.

CONSTRUCTION (DESIGN AND MANAGEMENT) REGULATIONS 2007

These regulations implement in Great Britain the requirements of Directive 92/57/EEC on minimum health and safety requirements at temporary or mobile construction sites. The regulations set out duties in respect of:

(a) the planning, management and monitoring of health, safety and welfare in construction projects; and
(b) the co-ordination of the performance of these duties by duty holders.

Projects

A *project* is defined as a project that includes or is intended to include construction work (as defined) and includes all planning, design, management or other work involved in a project until the end of the construction phase.

A project is notifiable to the HSE if the construction phase is likely to involve more than:

(a) 30 days or
(b) 500 person days

of construction work.

Duty holders

Five specific groups of people are identified as 'duty holders' under the regulations, namely:

Client – a person who in the course of a business or furtherance of a business:

(a) seeks or accepts the services of another who may be used in the carrying out of a project for him; or
(b) carries out a project himself.

CDM co-ordinator – a person appointed by a client, where a project is notifiable, to perform specified duties as per Regulations 20 and 21.

Designer – a person (including a client, contractor or other person referred to in the regulations) who in the course of furtherance of a business:

(a) prepares or modifies a design; or
(b) arranges for or instructs any person under his control to do so,

relating to a structure or to a product or mechanical or electrical system intended for a particular structure, and a person is deemed to prepare a design where a design is prepared by a person under his control.

Principal contractor – a person appointed by a client to perform the duties specified in Regulations 22 to 24 as soon as is practicable after the client knows enough about the project to be able to select a suitable person for such appointment.

Contractor – a person (including a client, principal contractor or other person referred to in the regulations) who, in the course of furtherance of a business, carries out or manages construction work.

Competence requirements

No person on whom these regulations place a duty shall:

(a) appoint or engage a CDM co-ordinator, designer, principal contractor or contractor unless he has taken reasonable steps to ensure that the person to be appointed or engaged is competent;
(b) accept such an appointment or engagement unless he is competent;
(c) arrange for or instruct a worker to carry out or manage design or construction work unless the worker is competent or under the supervision of a competent person.

Co-operation and co-ordination

There is a duty on all persons concerned with a project, at the same or an adjoining site, to co-operate with each other. They must further co-ordinate their activities in a manner that ensures, so far as is reasonably practicable, the health and safety of persons carrying out the construction work and those affected by the construction work.

General principles of prevention

Every duty holder shall, in relation to the design, planning and preparation of a project, and with respect to the construction phase of a project, take account of the general principles of prevention in the performance of those duties.

Clients' duties

A client has a duty to:

(a) take reasonable steps to ensure that the arrangements for managing the project are suitable; and
(b) ensure that every designer and every contractor is promptly provided with specified pre-construction information.

Designer's duties

A designer must:

(a) not commence work in relation to a project unless the client is aware of his duties under the regulations;
(b) avoid foreseeable risks to the health and safety of any person in preparing or modifying a design that may be used in construction work;
(c) eliminate hazards that may give rise to risks and reduce risks form any remaining hazards;
(d) in designing a structure for use as a workplace, take account of the provisions of the Workplace (Health, Safety and Welfare) Regulations 1992; and
(e) take all reasonable steps to provide with his design sufficient information about aspects of the design of the structure or its construction or maintenance as will adequately assist clients, other designers and contractors to comply with their duties under the regulations.

Principal contractor's duties

A principal contractor shall:

(a) plan, manage and monitor the construction phase in a way that ensures, so far as is reasonably practicable, that it is carried out without risk to health or safety, including facilitating co-operation and co-ordination between persons concerned and the application of the general principles of prevention;
(b) liaise with the CDM co-ordinator in relation to any design or change to a design;
(c) ensure sufficient welfare facilities are provided;
(d) where necessary, draw up rules that are appropriate to the construction site and the activities on it ('site rules');
(e) give reasonable directions to any contractor to enable the principal contractor to comply with his duties;

(f) ensure that every contractor is informed of the minimum amount of time that will be allowed to him for planning and preparation before he commences work;
(g) where necessary, consult a contractor before finalising such part of the construction phase plan as is relevant to the work to be performed by him;
(h) ensure that every contractor, prior to commencing construction work, is given access to such part of the construction phase plan as is relevant;
(i) ensure that every contractor is given such further information as he needs;
(j) identify to each contractor the information that is likely to be required by the CDM co-ordinator for inclusion in the health and safety file;
(k) ensure that the particulars required in the notification of a project are displayed in a readable condition; and
(l) take reasonable steps to prevent access by unauthorised persons to the construction site.

The principal contractor must take reasonable steps to ensure that every worker is provided with suitable site induction, appropriate information and training, and any further information and training for particular work to be carried out without undue risk to health or safety.

The principal contractor shall:

(a) before start of the construction phase, prepare a construction phase plan that is sufficient to ensure that the construction phase is planned, managed and monitored to enable construction work to commence without risk to health or safety;
(b) from time to time update, review, revise and refine the plan; and
(c) arrange for the plan to be implemented in a way that ensures the health and safety of those involved in construction work and all persons who may be affected by that work.

The principal contractor shall make and maintain arrangements for effective co-operation, consultation and the provision of information in matters relating to the safety, health and welfare of all persons involved.

Contractors' duties

A contractor must:

(a) not carry out construction work in relation to any project unless the client is aware of his duties under the regulations;
(b) plan, manage and monitor construction work in a way that ensures that, so far as is reasonably practicable, it is carried out without risk to health and safety;

(c) ensure that any contractor appointed or engaged in connection with the project is informed of the minimum amount of time that will be allowed to him for planning and preparation before he begins construction work; and
(d) provide every worker under his control with any information and training that he needs for the particular work to be carried out safely and without risk to health.

CDM co-ordinator's duties

A CDM co-ordinator shall:

(a) give suitable and sufficient advice and assistance to the client on undertaking the measures he needs to take to comply with the regulations;
(b) ensure that suitable arrangements are made and implemented for the co-ordination of health and safety measures during planning and preparation for the construction phase of the project;
(c) take all reasonable steps to identify and collect the pre-construction information;
(d) promptly provide in convenient form to designers and contractors such of the pre-construction information in his possession as is relevant to each;
(e) take all reasonable steps to ensure that designers comply with their duties;
(f) take all reasonable steps to ensure co-operation between designers and the principal contractor during the construction phase;
(g) prepare, where none exists, and otherwise review and update a record (*health and safety file*) containing information relating to the project that is likely to be needed during any subsequent construction work and, at the end of the construction phase, to pass that file to the client; and
(h) as soon as practicable after his appointment ensure that notice is given to the HSE containing the particulars specified in Schedule 1.

Site safety requirements

Part IV lays down specific provisions with respect to:

(a) safe places of work;
(b) site good order and site security;
(c) stability of structures;
(d) demolition or dismantling;
(e) the use of explosives;
(f) excavations;
(g) cofferdams and caissons;

(h) reports of inspection;
(i) energy distribution installations;
(j) prevention of drowning;
(k) traffic routes;
(l) vehicles;
(m) prevention of risk from fire, etc;
(n) emergency procedures;
(o) emergency routes and exits;
(p) fire detection and fire fighting;
(q) fresh air;
(r) temperature and weather protection; and
(s) lighting.

Notification of a project

The following particulars of a project must be notified to the HSE prior to the commencement of work:

1. Date of forwarding.
2. Exact address of the construction site.
3. Name of the local authority for where the site is located.
4. A brief description of the project and the construction work that it includes.
5. Contact details of the client.
6. Contact details of the CDM co-ordinator.
7. Contact details of the principal contractor.
8. Date planned for the start of the construction phase.
9. The time allowed by the client to the principal contractor for planning and reparation for construction work.
10. Planned duration of the construction phase.
11. Estimated maximum number of people at work on the construction site.
12. Planned number of contractors on the construction site.
13. Name and address of any contractor already appointed.
14. Name and address of any designer already engaged.
15. A declaration signed by or on behalf of the client that he is aware of his duties under the regulations.

ACCIDENTS IN CONSTRUCTION ACTIVITIES

Falling from a height is the most common type of fatal accident in the construction industry. Manual handling accidents, frequently resulting in hernias, back injuries, foot and hand injuries, are the most common type of non-fatal accident.

Other frequent types of accident include:

(a) ladder accidents (the one out:four up rule for ladders should always be followed);
(b) falls from working platforms, such as scaffolds and mobile platforms;
(c) falls of materials, such as bricks, roof tiles, etc on to people working below;
(d) falls from pitched roofs or through fragile roofs;
(e) falls through openings in flat roofs and floors;
(f) collapses of excavations, resulting in people being buried alive;
(g) accidents associated with site transport, eg dumper trucks, reversing lorries;
(h) use of machinery and powered hand tools; and
(i) accidents associated with bad housekeeping.

REGULATING THE CONTRACTOR

Contractors undertake a wide range of services, from window cleaning to minor repairs and improvements to buildings to large-scale construction projects on existing or 'green field' sites. In the majority of situations the client and main contractor are jointly responsible for ensuring standards of safety, health and welfare are maintained during construction operations. Only in a 'green field' site situation, before new premises are handed over to the occupier, does the main contractor have full control and, therefore, responsibility, for a site.

The presence of contractors on site for a lengthy period of time can, if not carefully planned, create numerous problems, in addition to the risk of accidents to staff and site employees. To prevent such problems arising, therefore, it is essential that any project is discussed with the main contractor prior to commencement. In particular, the site of operations should be clearly defined on a factory or premises plan, indicating specific access and egress points for workers and materials, parking areas, vehicular traffic routes, including those of fork-lift trucks, areas prohibited to contractor's staff, the location of fire emergency points and specific hazards which may exist, such as the presence of flammable liquid storage points. Furthermore, agreement should be reached on the use or otherwise of company amenities by the contractor's staff, including washing, shower and toilet facilities, catering arrangements, first aid stations and, possibly, arrangements for drying clothing. The contractor should be advised of any company processes or activities which may expose building workers to health and safety risks.

It is normal practice for the contractor to take out insurance cover to indemnify the client in respect of any negligence resulting in personal injury and / or death, or damage to property and plant, arising out of or in connection with the work.

Other aspects which will need agreement between the client and contractor include:

(a) circumstances surrounding use of the owner's equipment;
(b) procedures for reporting, recording and investigating accidents on site;
(c) the use of safe systems of work, including permits to work;
(d) the safe use of plant and machinery, including electrical equipment and the supply of electricity to the site;
(e) the prevention of noise nuisance to the inhabitants of the neighbourhood;
(f) fire protection procedures, including the need for safe welding operations;
(g) arrangements for the safe removal of dangerous substances and wastes, in particular asbestos lagging or other asbestos-based materials which may be incorporated in the buildings due for redevelopment;
(h) the supply and use of appropriate personal protective equipment by site workers;
(i) permitted routes for contractors' vehicles, parking areas, delivery points, vehicle safety standards and driving restrictions;
(j) procedures for site clearance on completion; and
(k) the procedures for site security.

Local management should also agree a form of safety monitoring covering contractor activities with the main contractor, including the right to dismiss from the site any contractor, subcontractor or individual who, in the eyes of the company, has behaved unsafely to the extent of exposing himself, other site workers or members of the occupier's workforce to risk or injury or, alternatively, caused an accident to take place as a result of his negligence or lack of safety precautions.

SAFETY METHOD STATEMENTS

A safety method statement is a formally written safe system of work, or series of integrating safe systems of work, agreed between client and main contractor or main contractor and subcontractor, and produced where work with a foreseeably high hazard content is to be undertaken.

It should specify the operations to be carried out on a stage-by-stage basis and indicate the precautions necessary to protect site operators, staff occupying the premises where the work is undertaken and members of the public who may be affected directly or indirectly by the work.

It may incorporate information and specific requirements stipulated by clients, employers, health and safety specialists, enforcement officers, the police, site surveyors and the manufacturers and suppliers of plant, equipment and substances used during the work. In certain cases it may identify training needs, eg under the COSHH Regulations, Electricity at Work Regulations 1989, or the use of competent persons or specially trained operators.

The use of method statements

A safety method statement is necessary to ensure safe working in activities involving:

(a) the use of hazardous substances in large quantities, eg toxic, corrosive, harmful, irritant, flammable substances;
(b) the use of explosives;
(c) lifting operations;
(d) potential fire risk situations;
(e) electrical hazards;
(f) the use of sources of radiation;
(g) the risk of dust explosions or the inhalation of hazardous dusts, gases, vapours, fumes, etc;
(h) certain types of excavation, particularly those adjacent to existing buildings;
(i) demolition work; and
(j) the removal of asbestos from buildings.

Contents of a method statement

The following features may be incorporated in a safety method statement:

(a) techniques to be used;
(b) access provisions;
(c) safeguarding of existing work locations and positions;
(d) structural stability requirements, eg shoring;
(e) procedures to ensure the safety of others, eg members of the public;
(f) health precautions, including the use of local exhaust ventilation systems and personal protective equipment;
(g) plant and equipment to be used;
(h) procedures to prevent area pollution;
(i) segregation of certain areas;
(j) procedures for the disposal of toxic wastes; and
(k) procedures to ensure compliance with specific legislation, eg Environmental Protection Act, COSHH Regulations, Control of Noise at Work Regulations 2005, Control of Asbestos at Work Regulations.

Work involving asbestos

In addition to the above recommendations, the following features should be incorporated in a specific safety method statement where work entails the removal or stripping of asbestos:

(a) the specific safe system of work to be operated;
(b) procedures for segregation and sealing of the asbestos stripping area;
(c) personal protective equipment requirements;
(d) welfare amenity provisions – hand washing, showers, separation of protective clothing from personal clothing, sanitation arrangements, catering facilities, drinking water provision;
(e) ventilation requirements and arrangements for the working area;
(f) personal hygiene requirements for operators;
(g) supervision requirements and arrangements;
(h) atmospheric monitoring procedures, including action to be taken following unsatisfactory results from air samples taken; and
(i) notification requirements under the Control of Asbestos at Work Regulations 2002.

The need for method statements

Under the CDM Regulations the need for contractors to produce safety method statements prior to high risk operations must be considered by the client and his planning supervisor in the pre-tender stage health and safety plan.

Method statements are an important feature of a construction phase health and safety plan prepared by a main contractor prior to the commencement of a project. In certain cases a standard form of safety method statement is agreed, used and signed by the main contractor as an indication of his intention to follow that particular safe system of work.

CONTRACTORS' REGULATIONS

Most large organisations who employ contractors of all types on a regular basis operate a form of contractors' regulations which lay down the ground rules for any organisation operating on their site. It is a condition of contract that contractors and subcontractors comply with such regulations at all times.

MAINTENANCE WORK

Maintenance work may be of a planned and routine nature. Other forms of maintenance may be required in crisis situations.

The principal hazards associated with maintenance work can be classified:

1. **Mechanical:** Machinery traps, entanglement, contact, ejection; unexpected start-up of machinery.
2. **Electrical:** Electrocution, shock, burns, fire.
3. **Pressure:** Unexpected pressure releases, explosion.
4. **Physical:** Extremes of temperature, noise, vibration; dust and fumes.
5. **Chemical:** Gases, fogs, mists, fumes, etc.
6. **Structural:** Obstructions, floor openings.
7. **Access:** Work at heights, work in confined spaces.

The following precautions should be considered:

1. Safe systems of work.
2. Permit to work systems.
3. Designation of competent persons.
4. Use of method statements.
5. Operation of company contractors' regulations.
6. Controlled areas.
7. Access control.
8. Information, instruction and training.
9. Supervision arrangements.
10. Signs, marking and labelling.
11. Personal protective equipment.

SUMMARY

1. The relationship between contractor, subcontractor and client should be appreciated in terms of both criminal and civil liability in the event of accidents in construction activities.

2. Criminal provisions relating to safety, health and welfare on construction sites are detailed in the various construction regulations.

3. The principal type of construction accident, frequently with fatal results, is a fall from a height. Hernias, foot, hand and back injuries are common.

4. Under the CDM Regulations clients, CDM co-ordinators, designers, principal contractors and contractors have specific responsibilities.

5. With construction projects, a health and safety plan must be prepared.

6. A conscious effort should be made by organisations to regulate the activities of contractors, commencing with careful pre-planning of the operation prior to work commencing.

7. Contractors should be required to produce and comply with safety method statements for identified highly hazardous activities.

8. All organisations should operate formal contractors' regulations aimed at establishing the ground rules for contracting activities of all types on their premises.

9. Maintenance work can expose employees and other persons at work to a range of hazards which should be carefully controlled by employers.

REFERENCES

Anderson, P W P (1989) *Safety Manual for Mechanical Plant Construction*, Kluwer, London

Armstrong, P T (1980) *Fundamentals of Construction Safety*, Hutchinson, London

Building Regulations 2000 (SI 2000 No 2531), The Stationery Office, London

Construction (Design and Management) Regulations 2007, The Stationery Office, London

Construction (Head Protection) Regulations 1989, HMSO, London

Health and Safety Executive (1995) *Guidance for Builders and Contractors – Health and Safety for Small Construction Sites*, HSE Books, Sudbury

Health and Safety Executive (undated) *Health and Safety in Demolition Work, Pts I–IV* (Guidance Note 19/1–4), HMSO, London

Health and Safety Executive (undated) *Roofwork: Prevention of falls* (Guidance Note GS 10), HMSO, London

Health and Safety Executive (undated) *Safe Erection of Structures; Pts I–IV* (Guidance Note GS 28/1–4), HMSO, London

International Labour Organisation (1982) *Safety and Health in Building and Civil Engineering Work*, ILO, Geneva

16

Mechanical Handling

The movement and handling of goods may be by manual or mechanical means. This chapter examines the various forms of mechanical handling and the safety requirements necessary for such operations. (The various aspects of manual handling operations are covered in Chapter 5.)

Clearly, wherever possible, mechanical systems should be used in preference to manual ones. The type and form of the handling system will vary considerably according to the form of the loads to be handled, their shape, size, weight, the distance of travel and frequency of movement. Factors, such as the form of storage system in operation, such as a drive-in pallet racking system, the general layout of the premises and external areas, and procedures for the receipt and despatch of raw materials and finished products respectively, will also determine the type of mechanical handling system used.

MECHANICAL HANDLING SYSTEMS

Generally, mechanical handling systems take the following three forms:

(a) conveyors, which move goods either on a horizontal plane or at inclined angles;
(b) elevators, which operate on a vertical or inclined angle basis; and
(c) mobile handling equipment, such as fork-lift trucks which, fundamentally, transfer loads from one point to another, both internally and externally.

Conveyors

Conveyors take a number of forms according to the materials they may be carrying – roller conveyors, slat conveyors, trough conveyors, chain conveyors, belt conveyors and screw conveyors. While individual forms of conveyor may incorporate hazards specific to themselves, the more general hazards associated with all types of conveyor are outlined below:

(a) 'nips' or traps between moving parts, eg between belt and drive, between driven and tension rollers (see Figure 16.1);
(b) traps between moving and fixed parts, eg between slats on a slat conveyor and the guide frame containing the conveyor;
(c) traps and nips created by a drive mechanism, eg between chains and sprockets on a chain conveyor (see Figure 16.2);
(d) traps created at transfer points between two conveyors, eg between a roller conveyor and belt conveyor (see Figure 16.3);
(e) hazards associated with sharp edges to, for instance, worn conveyor chains, which may be exposed.

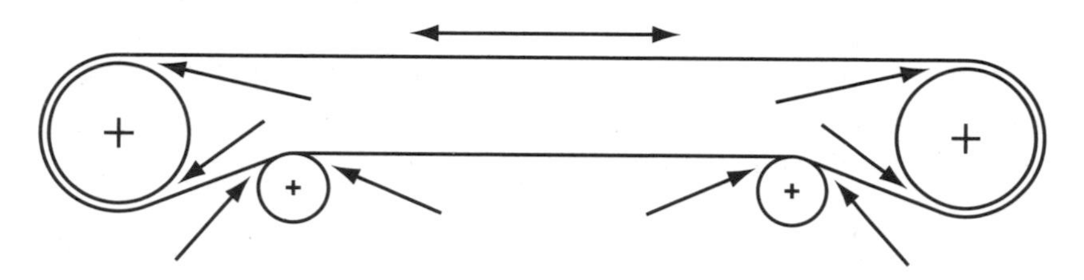

Figure 16.1 Trapping points on a reversible belt conveyor
(Source: BS EN ISO 12100: 2002 'Safety of Machinery')

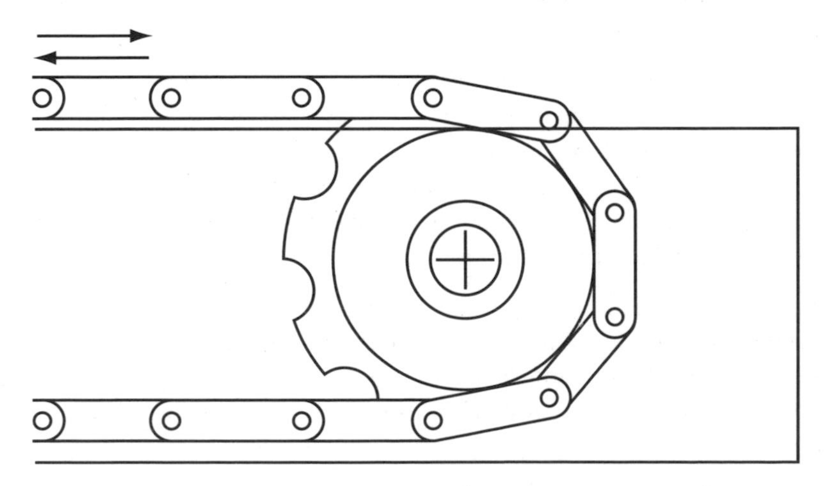

Figure 16.2 Traps between sprocket and chain on a reversible chain conveyor
(Source: BS EN ISO 12100: 2002 'Safety of Machinery')

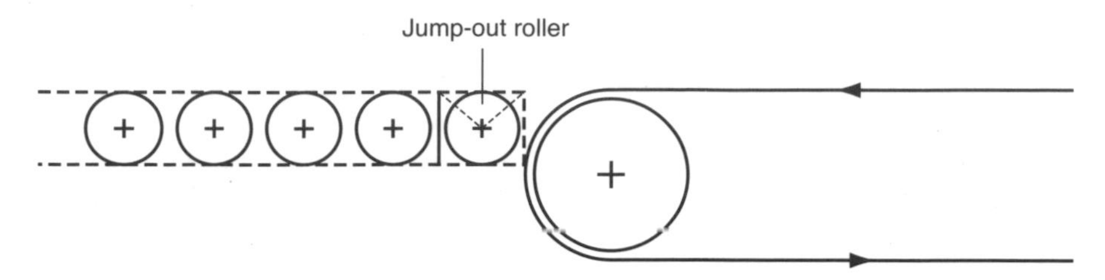

Figure 16.3 Safety requirements at the junction between a belt conveyor and roller conveyor

(Source: BS EN ISO 12100 'Safety of Machinery')

Guarding of conveyors

Fixed guards, which totally enclose the trap or nip created, should always be used in preference to other forms of guard. However, where frequent access may be needed to a danger point, eg for lubrication, an interlocked guard, which automatically stops the conveyor when it is lifted, should be installed. In certain cases, a tunnel guard, which is a form of distance guard, may be appropriate. (See Chapter 11.) This form of guard prevents access to the danger point because of the relationship of the guard opening dimensions to the length of the tunnel. Clearance between the sides of the opening of a tunnel guard and the goods being conveyed should be not less than 50 mm.

As with all other types of machinery, conveyor guarding should comply with BS EN ISO 12100 *Safety of Machinery*, particularly with regard to any openings in fixed guards.

Further points for consideration in the design and operation of conveyors include:

(a) arrangements for lubrication with the guards in position;
(b) maximising the radius of all bends to prevent items carried jamming on the conveyor or falling off same;
(c) elimination of sharp edges to supporting structural members; and
(d) the fitting of rails or side members to conveyors where the conveyor rises to more than 1 metre above floor level.

Where a conveyor is longer than 20 metres, an emergency stop (trip) wire should be fitted or, alternatively, a series of emergency stop buttons, spaced at not more than 10-metre intervals, should be provided. Stop buttons should be mushroom-shaped and red. They should remain in the 'OFF' position until reset.

Elevators

An elevator is used to transfer goods either vertically, eg between the floors of a building, or at a specific angle, eg between one floor and another. They may also be of the movable type for general use in warehouses and storage areas.

Both ends of an elevator should incorporate fixed guards due to the traps created between elevator chains and sprockets. Bucket elevators are commonly used for transferring loose materials, such as grain, inside a fixed shaft or hoistway. In this case there is a significant risk of dust explosions and, to prevent such explosions, all elevator heads must be fitted with explosion reliefs, usually a form of blow-out panel of a minimum size according to the size of the elevator.

With bar elevators, used for transferring sacked goods or boxed goods, traps can be created between the bar and the fixed part of the elevator. Such elevators should be guarded at the top and bottom by fixed guards.

Mobile work equipment

The most commonly used item of mobile work equipment is the fork-lift truck, which can include pedestrian-operated stacking trucks, reach trucks, counterbalance fork trucks, narrow aisle trucks and order pickers. Where such equipment is used, it is necessary to consider certain aspects such as the safe operation and maintenance of the equipment, specific provisions relating to the area of operation, and particular requirements relating to operators in terms of training, supervision and control, general fitness and the use of personal protective equipment. These aspects are dealt with below.

Mechanical handling equipment

To ensure safe operation, these rules should be followed:

(a) the maximum rated load capacity of the truck, as shown on the manufacturer's identification plate, should not be exceeded;
(b) passengers should never be carried, unless in a properly manufactured cage or platform;
(c) untrained and/or unauthorised personnel should never drive or operate such equipment;
(d) when driver-operated trucks are left unattended they should have the forks lowered and the truck should be immobilised by putting the controls in the neutral position, shutting off the power, applying the brakes, and removing the key or connector plug;
(e) the keys to the truck should be kept in a secure place when not in use;
(f) a formally established maintenance programme, based on the manufacturer's recommendations for inspection, maintenance and servicing, should be operated;

(g) drivers should be trained to undertake periodic maintenance checks, which should be linked to a formally established defect reporting system;
(h) trucks must comply with the Road Traffic Acts, in terms of lights, brakes, steering, etc when operating on a public highway; and
(i) all equipment used for mechanical handling and lifting should be subject to six-monthly and annual examinations. Lifting chains should be inspected annually and certificated as such in accordance with statutory requirements, ie all these items of equipment are classed as 'lifting machines' under section 27 of the FA, and the provisions relating to the examination and testing of same apply.

Area of operation

To ensure safe operation, consideration must be given to the design, layout and maintenance of operating areas. The following aspects must be considered:

(a) floors should be smooth, free from obstructions, changes in floor level and of adequate load-bearing capacity;
(b) sharp bends and overhead hazards, such as pipework, ducting and electrical conduits, should be eliminated;
(c) edges of loading bays should be protected when not in use, and tiger-striped floor markings incorporated at dangerous points to warn the driver;
(d) ramps and gradients should not exceed 10 per cent and there should be a gradual change of gradient at the bottom and the top of the slope;
(e) lighting should be maintained at a minimum level of 100 lux;
(f) bridge plates should be designed with an adequate safety margin to support loaded equipment;
(g) aisles should be wide enough, kept clear at all times and, in areas where other personnel are working on a spasmodic basis, fitted with photo-electrically operated audible warning devices, to warn such personnel of the presence of the truck;
(h) notices requiring drivers to sound their horns should be located at key points in the operating area, and corners should be fitted with convex mirrors to enable the driver to see people and goods prior to turning the corner;
(i) in truck charging areas, ventilation should be sufficient to prevent accumulations of hydrogen gas building up;
(j) refuelling of trucks should take place outside the area of operation; and
(k) truck parking areas should be located away from main vehicle parking areas.

Operator requirements

Operators of such equipment should be selected on the basis of their attitude to safe vehicle operation, physical fitness and attendance at an approved driver training course. Companies should operate a form of documentation of authorised operators through the use of a 'permit to drive' system. (An example of a company permit to drive is shown in Figure 16.4.)

Fork-lift truck handling operations, particularly in warehouses and stores where more than one truck may be in operation, require a high level of supervision and control, supported by regular training and retraining of operators. Operator training should be carried out by trainers who are experienced in the specific tasks to be undertaken.

PERMIT TO DRIVE

Name ____________________ Date of Birth ______________

Department ________________ Date of Basic Training _______

'Certificate No.' ____________

Permit Number ____________ Date of Issue ______________

The person named in this Permit is authorised to drive the following powered materials handling trucks:-

CLASS:– A; B1; B2; C; D; and E. (delete as appropriate)

Signature of Permit Holder ______________________________

Signature of Unit Manager _____________ Date ___________

CLASS OF TRUCK

A – pedestrian stacking;
B1 – pedestrian counterbalance;
B2 – rider counterbalance;
C – reach;
D – order picker;
E – narrow aisle.

NOTE

1 This permit does not authorise the person to whom it is issued to drive on a public road.
2 The permit is only valid for the class of truck specified.
3 This permit may be withdrawn if the safety rules laid down in the Alpha Company Code of Practice 'Materials Handling and Storage Equipment', are not observed.

Figure 16.4 Typical form of company permit to drive

Operator training should take place in the following three specific stages:

(a) acquisition of the specific skills and knowledge required to operate the equipment safely and to carry out the prescribed daily checks on the equipment;
(b) on-the-job training in a designated training area to develop operational skills; and
(c) familiarisation training, under close supervision of the trainer, in the workplace.

Further training should be carried out when:

(a) the driver is transferred to a new operational area or specific handling activity;
(b) new or modified equipment is introduced;
(c) there has been a significant change in the working layout; and
(d) there is evidence to indicate that driving standards have deteriorated or there has been a lapse in operator standards.

Operators should be provided with safety footwear and a safety helmet. In certain cases, the provision of hearing protection may be necessary. Protective clothing to suit weather and/or temperature conditions should also be provided where operations extend to external areas or into temperature-controlled parts of the premises, such as cold stores. In these cases donkey jackets and gloves should be provided.

A record should be maintained of all authorised operators. This record should indicate the:

(a) name of the operator;
(b) date of passing the truck driving test;
(c) serial number of the permit to drive;
(d) date of initial training and retraining; and
(e) classes of truck which the operator is authorised to drive.

Part III of PUWER lays down requirements for all forms of mobile work equipment. 'Mobile work equipment' is very broadly defined in the HSE guidance accompanying PUWER as 'any work equipment which carries out work while it is travelling or which travels between different locations where it is used to carry out work'. Such equipment would normally be moved on, for example, wheels, tracks, rollers or skids. It may be self-propelled, towed or remote controlled, and may incorporate attachments. Under Part III:

(a) Mobile work equipment must be suitable for carrying persons and incorporate features for reducing risks to their safety (Regulation 25).
(b) Measures must be taken to minimise risks from rolling over of mobile work equipment (Regulation 26).

(c) Fork-lift trucks must be adapted or equipped to reduce the risk from overturning (Regulation 27).
(d) Self-propelled work equipment must incorporate a number of safety features, including facilities for preventing it from being started by an unauthorised person (Regulation 28).
(e) Remote-controlled self-propelled work equipment must incorporate an automatic stop device which operates once the equipment leaves its control range (Regulation 29).
(f) There are specific safety provisions for drive shafts to mobile work equipment with respect to means for preventing seizure of a shaft and means for safeguarding a shaft against soiling or damage by contact with the ground while uncoupled (Regulation 30).

The HSE publication *Rider-operated Lift Trucks – Operator Training: Approved Code of Practice and Supplementary Guidance* (1988) provides excellent guidance on this matter.

SUMMARY

1. Conveyors can be exceedingly dangerous if not effectively guarded in accordance with BS EN ISO 12100 and there is an absolute duty under PUWER to guard same.
2. Fixed guards should always be used on conveyor danger points wherever practicable, together with the provision of emergency stop devices.
3. Elevators should be guarded at both ends due to traps created by chains and sprockets (toothed wheels) in particular.
4. Where elevators are used for the transfer of dust-forming materials, explosion reliefs should be installed at the head and the hoistway / shaft and floor openings constructed in materials to give a half-hour notional or predicted period of fire resistance.
5. If they are not properly controlled, mechanical handling equipment, such as fork-lift trucks, can represent a serious risk to the workforce.
6. Special consideration must be given to the equipment, the design and layout of the working area, and to the selection, training, supervision and control of vehicle operators.
7. Clearly established maintenance procedures should be operated in the case of all mobile handling equipment.
8. A formal permit to drive system should be operated with a view to regulating the selection and appointment of truck drivers. Where there is

evidence of unsafe driving or truck operation, consideration must be given to the withdrawal of the permit to drive as a form of disciplinary action.

REFERENCES

British Standards Institution (2002) *BS EN ISO 12100: Safety of Machinery*, BSI, Milton Keynes

Health and Safety Executive (1979) *Lift Trucks*, HMSO, London

Health and Safety Executive (1980) *Safe Working with Lift Trucks (Guidance Note HS(G) 6)*, HMSO, London

Health and Safety Executive (1988) *Approved Code of Practice: Rider-operated lift trucks – operator training*, HMSO, London

Health and Safety Executive (1998) *Provision and Use of Work Equipment Regulations 1998 and Guidance on Regulations*, HMSO, London

Royal Society for the Prevention of Accidents (1975) *Training Manual for Power Truck Operators*, RoSPA, Birmingham

17

The Working Environment

Section 2 of the HASAWA places a duty on the employer to provide and maintain a working environment that is safe and without risks to health, including arrangements for the welfare of employees while at work. The Workplace (Health, Safety and Welfare) Regulations extend this duty further.

As such, the concept of the working environment covers a very broad range of aspects that have to be considered, in particular the location and layout of working areas, the use of colour systems for waste disposal and traffic management, and the elimination or control of environmental stressors. Environmental stressors include extremes of temperature, lighting and ventilation, noise and vibration, and the presence of dusts and gases, all of which can have direct and indirect effects on the health of workers. The provision and maintenance of welfare amenity provisions are also important features of the working environment.

LOCATION AND LAYOUT OF WORKPLACES

In considering the possible location of a workplace, eg a factory, office block, warehouse, etc consideration should be given to factors such as ease of access and exit for vehicles and pedestrians, the availability of public transport and vehicle parking areas, the potential for nuisance being created to the inhabitants of the neighbourhood, the vulnerability of the general public to major incidents, such as large fires or multiple road accidents, and the density of surrounding buildings.

The duties of the employer towards non-employees are clearly detailed in Section 4 of the HASAWA.

The layout of a workplace is significant in terms of preventing overcrowding, ensuring orderly manufacture of the product and the maintenance of satisfactory safety standards.

Regulation 10 of the Workplace Regulations stipulates that 'every room where persons work shall have sufficient floor area, height and unoccupied floor space for the purposes of health and safety'. In workplaces established before these regulations came into force, ie 1 January 1993, the schedule to the regulations specifies a minimum of 11 cubic metres per person, with no space more than 4.2 metres from the floor being taken into account in any such calculation.

THE USE OF COLOUR IN THE WORKPLACE

Good choice of colour can promote a congenial environment which assists in achieving high standards of work performance. From the point of view of safety, the careful use of colour and colour schemes can do much to increase staff awareness of hazards and of the actions necessary should such hazards arise. Typical examples are the use of safety signs and symbols, as per the Safety Signs Regulations 1980, and in the yellow and black 'tiger striping' of floor and road surfaces to identify traffic hazards.

Careful choice of wall and ceiling colours also helps maintain good levels of illumination.

WASTE DISPOSAL

Provision must be made for the collection, storage and disposal of waste materials from premises. While most liquid wastes are removable through the drainage system, solid wastes, particularly those which are flammable, can present a problem. The law requires that no waste material or refuse should be allowed to accumulate within a working area and that an adequate supply of waste containers should be provided. Refuse and waste materials should be removed from the premises on a daily basis and stored in a designated waste storage area. Where bulky wastes are produced, the use of an industrial waste compactor linked to a bulk waste container is recommended, rather than standard dustbins.

The local authority environmental health and engineering departments should be consulted wherever toxic or dangerous wastes must be removed and disposed of at an approved disposal site. In the majority of cases, specialist waste disposal contractors will provide such a service.

ENVIRONMENTAL CONTROL

The following are the environmental elements that can be controlled.

Temperature

Exposure to extremes of temperature can result in heat stress, heat stroke, and at the other extreme, frostbite. Moreover, temperature control of the workplace, along with control over air movement and relative humidity, is important in the maintenance of comfort conditions.

Under the WHSWR, the temperature in all workplaces inside buildings shall be reasonable during working hours. (See Table 17.1.) No method of heating that exposes any person to the risk of exposure to injurious or offensive fumes, gas or vapour may be used, and a sufficient number of thermometers must be provided to enable persons at work to determine the temperature of an indoor workplace. Workplaces must be adequately thermally insulated where necessary, having regard to the type of work carried out and the physical activity of the persons carrying out the work. Excessive effects of sunlight on temperature must be avoided (Regulation 7).

Type of work	*Temperature range*
Sedentary and / or office work	19.4 to 22.8°C
Light work	15.5 to 20.0°C
Heavy work	12.8 to 15.6°C

Table 17.1 Comfort temperature ranges

Relative humidity

Relative humidity should be between 30 and 70 per cent. Below this range discomfort is produced due to the drying of the throat and nasal passages, whereas above that range people will experience a feeling of stuffiness.

Lighting

Two aspects need consideration in lighting design:

(a) the quantity of light required for a given task, measured in lux (see following Note); and
(b) the quality of the lighting with regard to its distribution, the avoidance of 'glare' conditions, colour rendition and brightness.

Note: The standard unit of 'illuminance', that is the quantity of light required for a given task or area, is the lux. This equals one lumen per square metre. This unit replaced the foot candle which equated to the number of lumens per square foot. The term 'lumen' is the unit of luminous flux or light flow, describing the quantity of light received by a surface or emitted by a source of light.

Lighting standards are outlined in HSE Guidance Note HS(G) 38 *Lighting at Work*. In this document a relationship is drawn between the average illuminance and the degree or extent of detail which needs to be seen in a particular situation or task. Average illuminance values and minimum measured illuminance values, both measured in lux (lx), are shown in Table 17.2. The minimum measured illuminance is the lowest illuminance permitted in the work area, taking the requirements of health and safety into consideration.

Attention must also be paid to the relationship between the lighting of the work area and adjacent areas, perhaps used for storage. Substantial differences in illuminance levels between such areas may produce visual discomfort and even affect safety levels where there is frequent movement of pedestrians and vehicles, such as fork-lift trucks. To reduce risks to both operators and vehicles, as well as possible visual discomfort, maximum illuminance ratios are therefore recommended in the Guidance Note. (See Table 17.3.)

Where there is conflict between the recommended average illuminances shown in Table 17.2 and the maximum ratios of illuminance in Table 17.3, the higher value should be taken as the appropriate average illuminance.

'Glare' is a problem which is occasionally encountered in lighting installations, and is the effect of light which causes discomfort or impaired vision. It is experienced when parts of the visual field are excessively bright compared with the general surroundings. It frequently occurs when the light source is directly in line with the task being undertaken or when light is reflected off a given object or surface. Glare can occur in the following three forms.

1. Disability glare

This is the visually disabling effect caused by bright bare lamps directly in the line of sight. The impaired vision (dazzle) which results from this effect may be hazardous in the driving situation, when working in high-risk operations or at heights.

2. Reflected glare

This is the reflection of bright light sources on wet or shiny work surfaces, such as plated metal or glass. The effect is to totally conceal the detail in or behind the object which is glinting. Wherever possible, light sources of low brightness levels should be used and the geometry of the lighting installation should be arranged so that there is no glint at the viewing position.

General activity	*Typical locations/types of work*	*Average illuminance (lx)*	*Minimum measured illuminance (lx)*
Movement of people, machines and vehicles (1)	Lorry parks, corridors, circulation routes	20	5
Movement of people, machines and vehicles in hazardous areas; rough work not requiring perception of detail	Construction site clearance, excavation and soil work, loading bays, bottling and canning plants	50	20
Work requiring limited perception of detail (2)	Kitchens, factories assembling large components, potteries	100	50
Work requiring perception of detail	Offices, sheet metal work, bookbinding	200	100
Work requiring perception of fine detail	Drawing offices, factories assembling electronic components, textile production	500	200

Table 17.2 Average illuminances and minimum measured illuminances
(Source: HSE Guidance Note HS(G)38 'Lighting at Work')

Notes

1. Only safety has been considered, because no perception of detail is needed and visual fatigue is unlikely. However, where it is necessary to see detail to recognise a hazard or where error in performing the task could put someone else at risk, for safety purposes as well as to avoid visual fatigue, the figure should be increased to that for work requiring the perception of detail.
2. The purpose is to avoid visual fatigue: the illuminances will be adequate for safety purposes.

3. Discomfort glare

This is caused by too much contrast of brightness between an object and its background. The phenomenon is generally associated with poor lighting design. Visual discomfort can result but the ability to see detail may not be impaired. Over a period of time operators exposed to discomfort glare can experience eye strain (visual fatigue), general fatigue and headaches.

Situations to which recommendation applies	*Typical location*	*Maximum ratio of illuminances*		
		Working area		*Adjacent area*
Where each task is individually lit and the area around the task is lit to a lower illuminance	Local lighting in an office	5	:	1
Where two working areas are adjacent, but one is lit to a lower illuminance than the other	Localised lighting in a works store	5	:	1
Where two working areas are lit to different illuminances and are separated by a barrier but there is frequent movement between them	A storage area inside a factory and a loading bay outside	10	:	1

Table 17.3 Maximum ratios of illuminance for adjacent areas
(Source: HSE Guidance Note HS(G) 38 'Lighting at Work')

The situation can be resolved in an existing work situation by:

(a) keeping luminaires as high as is practicable;
(b) maintaining luminaires parallel to the main direction of lighting; and
(c) careful design of shades which screen the lamp.

Reference should be made to the Illuminating Engineering Society's (IES) schedule of Limiting Glare Indices in the design or improvement of lighting installations. These are indices representing the degree of discomfort glare which will be just tolerable in the process or location under review. Where these indices are exceeded, occupants may experience visual fatigue and/or headaches.

Lighting distribution is another important feature of lighting design. Distribution is concerned with the way light is spread, and is classified under the British Zonal Method from BZ1 (all light downward in a vertical column) to BZ10 (light in all directions). For good general lighting, regularly spaced luminaires should be used to give an evenly distributed illuminance. This evenness of illuminance depends upon the ratio between the height of the luminaire above the working position and the spacing of fittings. The IES spacing:height ratio provides a general guide to such arrangements, the normal ratio being 1½:1 or 1:1, depending upon the type of luminaire.

Brightness (luminosity) is a subjective sensation and cannot be measured. However, a brightness ratio can be considered, which is the ratio of apparent luminosity between a task object and its surroundings. All surfaces have a particular level of reflectance, ie the ability of a surface to reflect light. Given a task illuminance factor of 1, the comparable reflectance values should be as shown in Table 17.4.

Ceilings	0.6
Walls	0.3 to 0.8
Floors	0.2 to 0.3

Table 17.4 Effective reflectance values

(Source: Illuminating Engineering Society)

Colour rendition refers to the appearance of an object under a given light source, compared to its colour under a reference illuminant, eg natural light, and enables the true colour to be correctly perceived.

Generally, the colour rendering properties of fitments should not be in stark contrast to those of natural light. They should also be as effective at night as during the day, as then there will be no daylight contribution to total workplace illumination.

Ventilation

Two aspects must be considered to ensure satisfactory ventilation of the workplace, namely the provision of sufficient air for people to breathe and work in reasonable comfort (comfort ventilation) and, secondly, means for the removal of air which may have become contaminated as a result of processes (exhaust/ extract ventilation). The first requirement is dealt with in the Workplace Regulations, whereas the requirement to provide effective exhaust ventilation is covered by the COSHH Regulations.

In the case of comfort ventilation, Regulation 6 of the Workplace Regulations states that:

1. Effective and suitable provision shall be made to ensure that every enclosed workplace is ventilated by a sufficient quantity of fresh or purified air.
2. Any plant used for the purpose of complying with the above paragraph shall include an effective device to give visible or audible warning of any failure of the plant where necessary for reasons of health or safety.

The requirements, therefore, are clear. There must be sufficient circulation of fresh air to maintain comfort conditions and a healthy working environment. Where there is emission of dust, fumes, gases and other forms of airborne pollution from processes then, additionally, effective systems of exhaust ventilation must be provided and maintained.

Mechanical ventilation systems

Reliance on natural ventilation, ie from doors, windows and other openings in the fabric of a workplace, is totally unsatisfactory where there may be emissions of dust and fumes from processes. In these cases mechanical ventilation systems must be operated. Such systems take two principal forms – extract (exhaust) ventilation and dilution ventilation.

Extract ventilation (exhaust ventilation): There are three main types of extract ventilation system – receptor systems, captor systems and low- volume high-velocity (LVHV) systems (see Figure 17.1).

1. *Receptor systems:* With this type of system the contaminant enters the system without inducement. The fan maintains a flow of air to transport the contaminant from its point of emission through a hood and ducting to a collection system. In some cases the hood actually forms a total enclosure around the source, eg a laboratory fume cupboard, or a partial enclosure, eg a paint spray booth.
2. *Captor systems:* As the term implies, moving air captures the contaminant at some point outside the receiving hood and induces its flow into same. The rate of air flow into the hood, which can take a number of different shapes, must be high enough to capture the contaminant at the furthest point of origin. Moreover, the air velocity induced at this point must be sufficient to prevent the contaminant flowing in another direction away from the hood. Captor systems are frequently used for welding activities, other than in a specifically designed welding booth or bay, with certain types of grinding machines and in soldering operations, the principal objective being to take the offending airborne contaminant away from the operator's breathing zone.
3. *Low-volume high-velocity (LVHV) systems:* With certain high-speed grinding machines or pneumatically operated chipping tools very high capture velocities are required. To prevent environmental contamination and the risk of dust being inhaled by the operator, LVHV systems are used. These systems operate on the principle of extracting the dust through small apertures very close to the point of emission, thereby achieving high velocities at the source and with low air flow rates.

Such systems need regular maintenance, together with comprehensive operator training.

RECEPTOR SYSTEMS

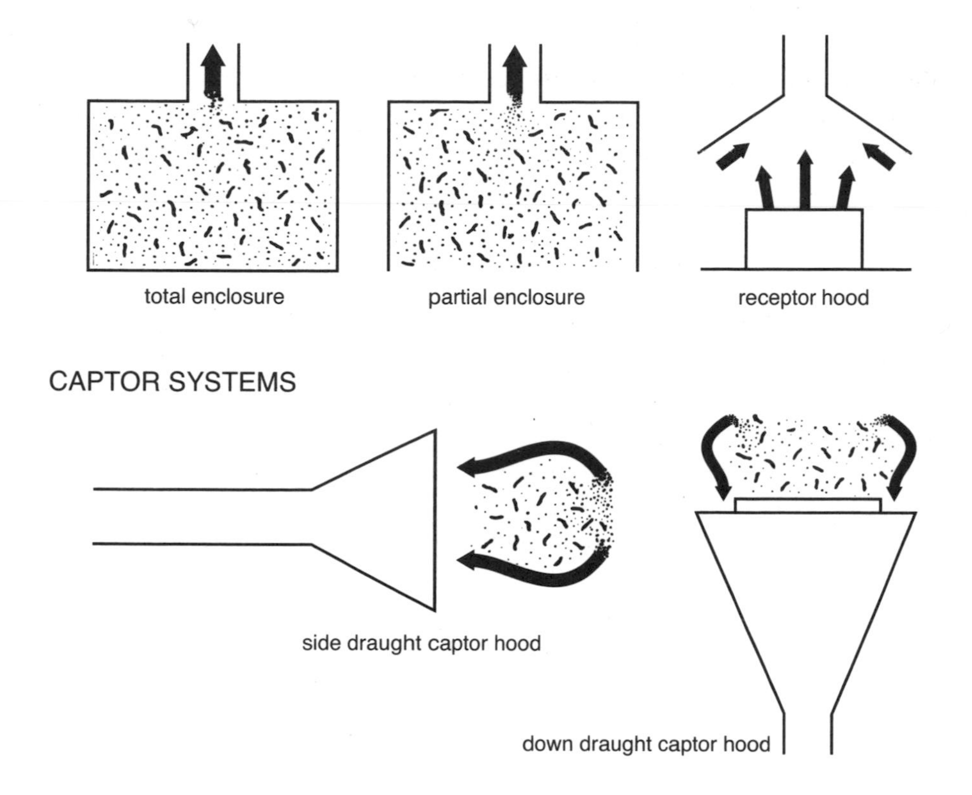

LOW-VOLUME HIGH-VELOCITY SYSTEMS

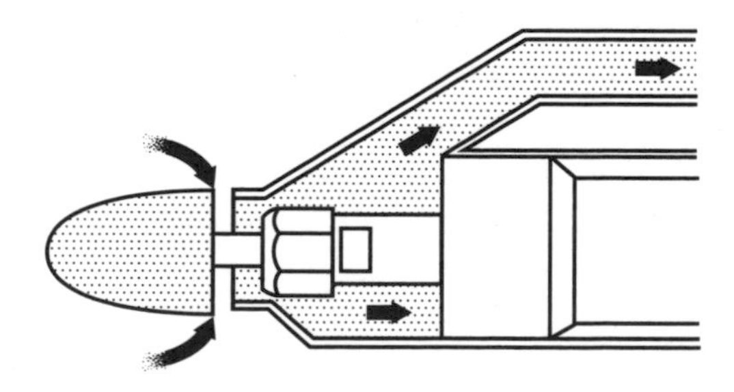

Figure 17.1 Extract ventilation systems

Dilution ventilation

Dilution ventilation implies diluting the concentration of an offending agent, perhaps in gaseous form, with large quantities of fresh air to a point where the gas or vapour is no longer dangerous. This system is most successfully used to control vapours from low toxicity solvents in small quantities which are uniformly evolved from a process. Dilution ventilation is not recommended as a means of controlling dust and fume emission.

Ventilation and the COSHH Regulations

Under the COSHH Regulations an employer must either prevent or control exposure to hazardous substances. He must also ensure the controls are properly used, and maintain, examine and test these control measures. Local exhaust ventilation (LEV) systems, ie receptor systems and captor systems, are the most common form of control for airborne contaminants.

Examination and testing may be by means of visual checks, inspections, testing and servicing with a view to ensuring the system is maintained in effective working order. Manufacturers and suppliers of LEV systems should stipulate the method of testing and, in many cases, may undertake such testing on behalf of employers. In addition to effective preventive maintenance, there are statutory requirements to undertake formal examination and testing of LEV systems in certain industries, eg ceramics. Records of examination and tests should be kept, and must be available for inspection by employees, Employment Medical Advisers and inspectors.

NOISE

Noise is generally defined as 'unwanted sound'. Sources of noise in the workplace include:

(a) noise taking a structure-borne pathway;
(b) noise produced as a result of vibration in machines;
(c) radiation of structural vibration into the air;
(d) turbulence created by air or gas flow;
(e) noise taking an airborne pathway; and
(f) noise produced by vibratory hand tools, such as hand-held grinders and chain saws.

Exposure to noise

Exposure to noise may affect hearing in three ways:

(a) *Temporary threshold shift* is the short-term effect, ie a temporary reduction in hearing acuity, which may follow exposure to noise. The condition is reversible and the effect depends to some extent on individual susceptibility.
(b) *Permanent threshold shift* takes place when the limit of tolerance is exceeded in terms of time, the level of noise and individual susceptibility.

Recovery from permanent threshold shift will not proceed to completion, but will effectively cease at some particular point in time after the end of the exposure.

(c) *Acoustic trauma* is quite a different condition from occupational deafness (noise-induced hearing loss). It involves sudden aural damage resulting from short-term intense exposure or even from one single exposure. Explosive pressure rises are often responsible, such as exposure to gunfire, major explosions or even fireworks.

Protection strategies

People who are exposed to noise in excess of 85 dBA on a continuous or intermittent basis stand the risk of going deaf, ie noise-induced hearing loss or occupational deafness. (The decibel (dB) is a unit of sound pressure. Sound pressure level is a measurement of the magnitude of the air pressure variations or fluctuations which make up a particular sound. dB(A) implies measurement of sound pressure level using the 'A' network of a sound pressure level meter. The 'A' network of such an instrument is the one which most closely follows the performance of the human ear.)

Protection strategies directed at preventing workers from sustaining noise-induced hearing loss are, in order of importance, reduction of the noise at source, ie designing quieter machines, isolation of the noise source, ie by the use of soundproof enclosures and close shields, provision and use of hearing protection for workers exposed to noise, and reduction of the time during which people are exposed to noise.

It should be appreciated that the decibel scale is a logarithmic scale (not a linear scale) and, on this basis, each increase in pressure of three dBA represents a doubling of the sound intensity. Therefore, to ensure that operators do not receive more than a 90 dB dose of noise per eight-hour working day, sound pressure level should be related to the duration of exposure.

Noise control

Noise control implies, firstly, a consideration of the source of the noise (in machinery there may be many sources of noise) and, secondly, of the pathway taken by the noise to the recipient. The use of personal protection – ear muffs, acoustic wool, ear plugs – relies heavily on the exposed person using this equipment while exposed to the particular noise level and, as such, will never be a perfect solution to the problem.

Noise control strategies should take into account the sources, pathways and control measures.

CONTROL OF NOISE AT WORK REGULATIONS 2005

These regulations specify *exposure action values* and *exposure limit values*. These values are:

- *lower exposure action values (LEAV):*
  (i) a daily or weekly personal noise exposure of 80 dBA; or
  (ii) a peak sound pressure of 135 dBC.
- *upper exposure action value (UEAV):*
  (i) a daily or weekly personal noise exposure of 85 dBA; or
  (ii) a peak sound pressure of 137 dBC.
- *an exposure limit value (ELV):*
  (i) a daily or weekly personal noise exposure of 87 dBA; or
  (ii) a peak sound pressure of 140 dBC.

Principal duties of employers

Where work is liable to expose employees to noise at or above a *lower exposure action value*, an employer shall make a suitable and sufficient assessment of the risk from the noise to the health and safety of employees.

An employer:

(a) shall ensure that employees are not exposed to noise above an *exposure limit value* (ELV); or
(b) if an ELV is exceeded, forthwith:
  (i) reduce exposure to noise to below the ELV;
  (ii) identify the reason for that ELV being exceeded; and
  (iii) modify the organisational and technical measures taken so as to prevent it being exceeded again.

Risk assessment

The risk assessment must identify the measures that need to be taken to comply with the regulations and must take into account certain specified aspects, such as the level, type and duration of exposure, including any exposure to peak sound pressure. The employer must further consult the employees concerned or their representatives on the assessment of risk.

Assessment of noise shall be by means of:

(a) observation of specific work practices;
(b) reference to relevant information on probable levels of noise corresponding to any equipment used; and
(c) if necessary, measurement of the level of noise to which employees are likely to be exposed.

Prevention or control of exposure

An employer must ensure that the risk is either eliminated at source or, where not reasonably practicable, to as low a level as is reasonably practicable. If an employee is likely to be exposed to noise at or above a UEAV, the employer must reduce exposure to as low a level as is reasonably practicable by establishing and implementing a programme of organisational and technical measures, *excluding* the provision of hearing protectors, that is appropriate to that activity.

General principles of prevention

An employer must ensure that the action taken is based on the general principles of prevention set out in Schedule 1 to the Management of Health and Safety at Work Regulations (see page 36) and shall include consideration of:

(a) other working methods that reduce exposure to noise;
(b) choice of appropriate work equipment emitting the least possible noise, taking account of the work to be done;
(c) the design and layout of workplaces, workstations and rest facilities;
(d) suitable and sufficient information for training of employees, such that the work equipment may be used correctly, in order to minimize exposure to noise;
(e) reduction of noise by technical means;
(f) appropriate maintenance programmes for work equipment, the workplace and workplace systems;
(g) limitation of the duration and intensity of exposure to noise; and
(h) appropriate work schedules with adequate rest periods.

Support strategies

An employer must:

(a) make personal hearing protection available:
    (i) upon request, to any employee who carries out work that is likely to expose him to noise at or above a LEAV;
    (ii) to an employee so exposed where reducing noise levels to below a UEAV by other means cannot be accomplished;
(b) designate hearing protection zones in cases where an employee is likely to be exposed to noise at or above a UEAV;
(c) where a risk assessment identifies a hearing risk, ensure that such employees are placed under suitable health surveillance;
(d) provide suitable information, instruction and training (which must incorporate specified training topics) where employees are likely to be exposed to noise at or above the LEAV.

Employees' duties

Employees must:

(a) make full and proper use of hearing protection and other control measures;
(b) report defects in hearing protection and other control measures; and
(c) present themselves for health surveillance procedures where they are found to have identifiable hearing damage.

VIBRATION-INDUCED INJURY

Exposure to vibration can have significant long-term effects on the body. This may arise as a result of driving large vehicles on uneven terrain and in the extensive use of certain types of hand-held power tools.

Legal requirements covering these risks are outlined in the Control of Vibration at Work Regulations 2005.

- *Whole body vibration (WBV)*
  The vibration of the body while driving heavy goods vehicles for long distances or on rough ground during, for instance, construction operations, can affect people in several ways. Drivers have experienced loss of balance, loss of concentration and blurred vision, in some cases a contributory factor in vehicle accidents.
  In the longitudinal direction ie head to feet, the human body is most sensitive in the range 4 to 8 Hz. In the transverse direction, ie finger tip to finger tip, the body is most sensitive in the 1 to 2 Hz range.
- *Hand-arm vibration syndrome (HAVS)*
  This condition arises through regular exposure to vibration from the use of certain types of hand tools. Exposure can cause damage to the circulatory system. Typical symptoms are parasthesia ('pins and needles') in the fingers, a reduced sense of touch and strength of grip, together with severe pain and numbness in the arm, wrist and hand.
- *Vibration-induced white finger (VWF)*
  This condition is well known in many industries where employees use certain forms of vibratory hand tools. The first signs of VWF are a mild tingling and numbness in the fingers. As exposure continues, the tips of the fingers become white, particularly during adverse weather and early in the morning. Eventually, this blanching increases to the full extent of the fingers and, on continuing exposure, the fingers take on a blue-black appearance.
  Early treatment is vital as VWF can progress to gangrene and necrosis (death of tissue) resulting in the need for surgical amputation of parts of fingers or whole fingers.

CONTROL OF VIBRATION AT WORK REGULATIONS 2005

Duties of employers

- Employers have the duty to undertake vibration risk assessments, taking into account the specified factors to be considered in such a risk assessment, where there is risk of injury arising from hand-arm vibration or whole-body vibration.
- They must take into account:
  - Exposure limit values – a level of daily exposure for any worker which must not be exceeded.
  - Exposure action values – the level of daily exposure for any worker which, if exceeded, requires specific action to be taken to reduce risk.
- Exposure action values and exposure limit values are specified thus:
  - for hand-arm vibration:
    - The daily exposure limit value normalised to an 8-hour reference period is 5 m/s2.
    - The daily exposure action value normalised to an 8-hour reference period is 2.5 m/s2.
  - for whole-body vibration:
    - The daily exposure limit value normalised to an 8-hour reference period is 1.15 m/s2.
    - The daily exposure action value normalised to an 8-hour reference period is 0.5 m/s2.
- Elimination or control of exposure is based on 'Principles of Prevention' set out in Schedule 1 to the Management of Health and Safety at Work Regulations.
- Where it is not reasonably practicable to eliminate at source and the risk assessment indicates an exposure action value is likely to be exceeded, an employer must establish a programme of organisational and technical measures appropriate to the activity and consistent with the risk assessment.

WORK INVOLVING RADIOACTIVE SOURCES

Radiation is a form of energy that involves the process of ionisation of atoms, such as hydrogen. This energy is released in the form of waves that can affect the human body in different ways.

Ionisation

An ion is a charged atom or group of atoms. Where the number of electrons does not equal the number of protons, the atom has a net positive or negative charge and is said to be 'ionised'. If a neutral atom loses an electron, a positively charged ion will result. Thus, ionisation is the process of losing or gaining electrons and occurs in the course of many physical and chemical reactions.

Hazards arising from exposure to ionising radiation

The effects of a dose of ionising radiation vary according to the type of exposure, for instance, whether the dose was local, affecting only a part of the body surface, or general, affecting the whole body. Furthermore the actual duration or length of time of exposure determines the severity of the outcome of such exposure.

1. Local exposure

This is the most common form of exposure and may result in reddening of the skin with ulceration in serious cases. Where exposure is local and the dose small, but of long duration, loss of hair, atrophy and fibrosis of the skin can occur.

2. General exposure

The effects of acute general exposure range from mild nausea to severe illness, with vomiting, diarrhoea, collapse and eventual death. General exposure to small doses may result in chronic anaemia and leukaemia. The ovaries and testes are particularly vulnerable and there is evidence that exposure to radiation reduces fertility and causes sterility.

Apart from the danger of increased susceptibility to cancer, radiation can damage the genetic structure of reproductive cells, causing increases in the number of stillbirths and malformations.

Biological dose of radiation

The unit of biological dose of radiation is the Sievert. Specified dose limits are dealt with in the Ionising Radiations Regulations 1999.

Radiological protection

A significant feature of radiological protection is the form taken by the radioactive source. Sources may be sealed or unsealed.

1. Sealed sources of radiation

The source is contained in such a way that the radioactive material cannot be released, as with X-ray machines. The source of radiation can be a piece of radioactive material, such as cobalt, which is sealed in a container or held in another material that is not radioactive. It is usually solid and the container and any bonding material are regarded as the source.

2. Unsealed sources of radiation

Unsealed sources may take many forms, such as gases, liquids and particulates. As they are unsealed, entry into the body is comparatively easy.

Criteria for radiological protection

These are based on three specific factors – time, distance and shielding. The principal objective is to ensure no one receives a harmful dose of radiation.

1. Time

Radiation workers may be protected by limiting the duration of exposure to certain pre-determined time limits.

2. Distance

Workers can be protected by ensuring they do not come within certain distances of radiation sources. This may be achieved by the use of restricted areas, barriers and similar controls. The Inverse Square Law applies in this case.

3. Shielding

Persons may be shielded by the use of absorbing material, such as lead or concrete, between themselves and the source to reduce the level of radiation to below the maximum dose level. The quality and quantity (thickness) of shielding varies for the radiation type and energy level and varies from no shielding through lightweight shielding, for instance, 1 centimetre-thick Perspex, to heavy shielding, such as centimetres of lead or metres of concrete.

Radiological protection procedures

1. Pre-employment and follow-up medical examinations.
2. Appointment of qualified persons (radiation protection advisers).
3. Maintenance of individual dose records.
4. Information, instruction and training and hazard awareness.
5. Continuous and spot-check radiation (dose) monitoring (dosemeters, film badges) for classified persons.
6. Use of warning notices, controlled areas, supervised areas.
7. Strict adherence to maximum dose limits.

Additionally for unsealed sources:

8. Appropriate personal protective clothing and equipment.
9. Effective ventilation.

10. Enclosure / containment to prevent leakage.
11. Use of impervious surfaces.
12. Immaculate working techniques.
13. Use of remote control systems.

The central idea is the avoidance of radioactive contamination.

Ionising Radiations Regulations 1999

These regulations impose duties on employers to protect employees and other persons against ionising radiation arising fom work with radioactive substances and other sources of ionising radiation. Certain duties are imposed on employees.

The principal requirements of these regulations are:

1. Certain specified practices are prohibited without the authorisation of the HSE and specified work with ionising radiation must be notified to the HSE (Regulations 5 and 6).
2. Employers must make a prior assessment of the risks arising from their work with radiation, an assessment of the hazards likely to arise from that work, and prevent and limit the consequences of identifiable radiation accidents (Regulation 7).
3. Employers must take all necessary steps to restrict so far as is reasonably practicable the extent to which employees and other persons are exposed to ionising radiation (Regulation 8).
4. Respiratory protective equipment must conform to agreed standards, and personal protective equipment and other controls must be regularly examined and properly maintained (Regulations 9 and 10).
5. Limits of doses of ionising radiation which people may receive per calendar year are:

   20 millisieverts for employees of 18 years of age or above;
   6 millisieverts for trainees aged under 18 years;
   1 millisievert for any other person, including members of the public (Schedule 4).

6. In certain circumstances, employers must prepare contingency plans for radiation accidents (Regulation 12).
7. Employers must consult radiation protection advisers in respect of certain matters (Schedule 5) and provide adequate information, instruction and training to employees and other persons (Regulations 13 and 14).
8. Employers must designate controlled or supervised areas where there may be a need to restrict exposure, or where employees are likely to receive more than specified doses of radiation, set out appropriate local rules for

these areas, appoint radiation protection supervisors and monitor radiation levels in these areas (Regulations 16 to 19).

9. Where employees are likely to receive more than specified doses of ionising radiation they must be designated as classified persons, doses received by such persons must be assessed by an HSE-approved dosimetry service, and appropriate dose records maintained (Regulations 20 to 26).
10. Radioactive substances used as a source of ionising radiation must, whenever reasonably practicable, be in the form of a sealed source (Regulation 27).
11. Employers must make specific arrangements for the control of radioactive substances, articles and equipment (Regulations 28 to 33).

In the case of employees, they must not knowingly expose themselves or other persons to ionising radiation to an extent greater than is reasonably necessary, and must exercise reasonable care while carrying out such work (Regulation 34).

WELFARE AMENITY PROVISIONS

Amenities provided for workers include arrangements for sanitation (toilets, urinals), washing facilities and showers, the provision of drinking water and meals, and the storage and drying of clothing. Specific requirements are detailed in the Workplace Regulations (Regulations 21 to 25) and ACOP. Such requirements cover the number of water closets, urinals and wash basins on the basis of the number of people employed of both sexes, provisions relating to the accessibility of same, lighting and ventilation of amenity areas, and maintenance requirements. In certain industries controlled by specific regulations, eg Control of Lead at Work Regulations 2002, more specific provisions are laid down.

Welfare amenity provisions are dealt with principally in the Workplace Regulations. The principal requirements are outlined below, with more detailed information to be found in the ACOP to the regulations:

1. **Sanitary conveniences**

   Sanitary conveniences must be:

   (a) suitable and sufficient in number; one water closet must be provided for every 25 male and female employees, and fraction of 25 accordingly;
   (b) readily accessible;
   (c) adequately ventilated and lit;
   (d) kept in a clean and orderly condition; and
   (e) with separation of the sexes.

2. **Washing facilities**

Washing facilities must be:

(a) suitable and sufficient in number, with showers provided if necessary;
(b) adjacent to sanitary accommodation and changing rooms;
(c) provided with hot and cold water, soap, towels (or other suitable means of drying);
(d) adequately ventilated and lit;
(e) kept in a clean and orderly condition; and
(f) with separation of the sexes.

3. **Drinking water**

An adequate supply of wholesome drinking water must be provided in the workplace:

(a) which is readily accessible;
(b) conspicuously marked; and
(c) with the provision of cups or other drinking vessels, unless the supply is from a jet.

4. **Accommodation for clothing**

There must be provided:

(a) suitable and sufficient accommodation for personal clothing not worn during working hours and for special work clothing;
(b) adequate security for personal clothing;
(c) where necessary, separation of personal clothing from clothing worn at work;
(d) facilities for drying clothing; and
(e) a suitable location for the accommodation of clothing.

5. **Facilities for changing clothing**

Suitable and sufficient facilities for changing clothing must be provided where:

(a) the person has to wear special clothing for work; and
(b) the person can not, for reasons of health or propriety, be expected to change in another room.

Separate facilities, or the separate use of facilities, must be provided for men and women where necessary for reasons of propriety.

6. **Facilities for rest and to eat meals**

Suitable and sufficient rest facilities must be provided at readily accessible places. Rest facilities:

(a) where necessary for reasons of health or safety, must be provided in one or more rest rooms or, in other cases, in rest rooms or rest areas; and

(b) include suitable facilities to eat meals where food eaten in the workplace would otherwise be likely to be contaminated.

Suitable facilities must be provided for any person at work who is a pregnant woman or nursing mother to rest.

Suitable and sufficient facilities must be provided for persons at work to eat meals where meals are regularly eaten in the workplace.

New provisions relating to rest rooms and rest areas require that they shall be equipped with:

(a) an adequate number of tables and adequate seating with backs for the number of persons at work likely to use them at any one time; and
(b) seating that is adequate for the number of disabled persons at work and suitable for them (Regulation 25).

Where necessary, those parts of the workplace (including in particular doors, passageways, stairs, showers, washbasins, lavatories and workstations) used or occupied directly by disabled persons at work shall be organised to take account of their needs (Regulation 25).

SUMMARY

1. There is a legal duty on the employer under the HASAWA to provide and maintain a working environment that is safe and without risks to health, including arrangements for the welfare of employees while at work.
2. Control of the working environment should take into account the location and layout of the workplace, including means of access and exit, the prevention of overcrowding, the use of colour, systems for waste disposal and the need for sound traffic-management procedures.
3. Environmental stress can result in poor levels of operator performance, accidents and occupational ill health.
4. To prevent environmental stress, attention must be paid to systems for temperature and humidity control, lighting arrangements, ventilation, control over noise and vibration, and the removal of dangerous airborne contaminants.
5. Good standards of welfare amenity provision are essential features of the working environment.
6. Where work entails the use of radioactive sources, special precautions are required.

REFERENCES

Bilsom International (undated) *In Defence of Hearing*, Bilsom International Ltd, Henley-on-Thames
Bruel & Kjaer (1984) *Measuring Sound*, Bruel & Kjaer, Naerum, Denmark
Burns, W (1973) *Noise and Man*, John Murray, London
Electricity Council *Better Office Lighting*, Electricity Council, London
Health and Safety Commission (1992) *Workplace (Health, Safety and Welfare) Regulations 1992 and Approved Code of Practice*, HMSO, London
Health and Safety Executive (1975) *Principles of Local Exhaust Ventilation*, HMSO, London
Health and Safety Executive (1987) *Lighting at Work (Guidance Note HS(G) 38)*, HMSO, London
Health and Safety Executive (1991) *Noise at Work: Advice for employees*, HSE Information Centre, Sheffield
Health and Safety Executive (1992) *Listen Up*, HSE Information Centre, Sheffield
Health and Safety Executive (1999) *Ionising Radiations Regulations 1999 (SI 1999 No 3232)*, HMSO, London
Health and Safety Executive (2001) *Keep the Noise Down: Advice for purchasers of workplace machinery*, HSE Books, Sudbury
Health and Safety Executive (2005) *Control of Noise at Work Regulations 2005*, The Stationery Office, London
Health and Safety Executive (2005): *Control of Vibration at Work Regulations 2005*, The Stationery Office, London
Health and Safety Executive (undated) *Ventilation of the Workplace (Guidance Note EH/22)*, HMSO, London
Lyons, S (1984) *Management Guide to Modern Industrial Lighting*, Butterworths, Sevenoaks

18

Safety in Offices, Workshops and in Catering Operations

Most people would consider the average office to be a reasonably safe workplace compared, say, with a construction site or foundry. However, while fatal accidents are uncommon, minor accidents caused by trips and falls, the unsafe use of electricity, obstructed passages and stairways, failure to use access equipment and poor housekeeping are common.

Fire is, by far, the greatest hazard in offices. Because of the design, location, construction, layout and age of many offices, a fire can spread rapidly from floor to floor. The need for well-developed fire and emergency procedures in offices cannot be over-emphasised. This is particularly appropriate in large multi-occupied offices housed in old converted buildings.

SAFETY REQUIREMENTS FOR OFFICES

The following areas should be paid attention to regarding safety in offices.

Housekeeping

One of the greatest causes of accidents in offices, particularly falls, is that of bad housekeeping. The term 'housekeeping' implies 'everything in the correct place and a place for everything'. Bad housekeeping is frequently the cause of office fires, which can be caused by general untidiness and poor storage of flammable wastes, such as waste paper. The following housekeeping rules should be applied in offices:

(a) Keep the work area tidy. Items that are not in use should be stored away.
(b) Waste should be stored in waste containers, which should be emptied on a regular daily basis.
(c) Heavy items, such as ledgers, should not be placed on the top of cabinets or cupboards as they could fall on to someone using the cupboard.
(d) Harmful items, such as broken crockery, light bulbs and milk bottles, should be separately stored and disposed of.
(e) Passages, stairways, entrances and exits, in particular emergency exits, should be kept clear and free from surplus stationery, sacks of office refuse, surplus office equipment and furniture, and other large items.
(f) Spillages should be cleared up immediately.
(g) Damaged floor coverings, such as carpets, should be replaced immediately due to the tripping or slipping hazard created.
(h) An established cleaning schedule should be maintained.

Furniture and fittings

Bad office planning and layout with regard to the siting of furniture and fittings can result in numerous minor accidents. The following points should be considered with regard to office layout and the location of items such as filing cabinets, desks and equipment:

(a) Furniture should be arranged so that employees can move freely within the office and from one office to another.
(b) Doors and drawers, particularly to filing cabinets, should be kept closed when not in use.
(c) When using a filing cabinet, only one drawer should be opened at a time due to the risk of the cabinet tipping forward, particularly when the two top drawers are open at the same time.
(d) Filing cabinets should not be overloaded. Heavy items should be stored in the bottom drawer.
(e) The bottom drawer of a filing cabinet should always be closed immediately to avoid the risk of people tripping over it.
(f) Damaged and broken furniture and fittings, such as shelves, chairs and filing cabinets, should be repaired or replaced immediately.

Electrical appliances

Misuse and abuse of electricity in offices is one of the most significant causes of office fires. There should be a total prohibition on staff undertaking electrical repairs or modifications. The increased use of electrical appliances in offices has also contributed to the risk of electrical overloading. The following rules should be observed:

(a) Only trained and competent staff should attempt to repair electrically operated machinery.
(b) Machines should be switched off from the mains when left unattended for long periods.
(c) Cables should be so positioned that they do not trip people up. Where this is not possible, suitable permanent cable covers should be installed. Flexes should be shortened so that they do not trail across the floor or under desks.
(d) The use of freestanding radiant-type electric fires, sometimes used to supplement central heating in the winter months, should be prohibited.
(e) Cables to electrical appliances should be maintained in a sound condition and replaced when they become frayed or damaged.
(f) The use of multi-point adaptors should be either prohibited or carefully controlled in order to avoid overloading sockets.
(g) Electrical appliances should be examined every six months and a record of such examinations maintained.

Lifting and carrying

A substantial number of permanent back injuries are sustained by office staff lifting and carrying heavy items, such as electric typewriters, stationery packages and office furniture. To prevent the risk of such injuries, adequate manual-handling equipment, such as trolleys and sack trucks, should be provided. The general rule must be that no one should lift anything which is likely to cause injuries to the back, hands, arms, legs or feet.

Dangerous substances

A variety of products used in offices on a daily basis can be potentially dangerous. Such products include cleaning fluids, adhesives, quick drying inks and correcting fluids, all of which emit powerful fumes or vapours. In certain circumstances, such substances can be highly flammable. Other flammable substances include floor polishes and some aerosol-based cleaning compounds. To reduce the risks associated with dangerous substances, the following precautions are necessary:

(a) Staff should always read the manufacturer's instructions prior to using a potentially dangerous product.
(b) In certain situations it may be necessary to wear personal protective equipment, such as gloves, apron and goggles, when dealing with substances.

(c) Staff should be aware of the hazard warning symbols shown on the packages for potentially dangerous substances, eg flammable, toxic symbols.
(d) Waste should be disposed of safely. This particularly applies to cloths soaked in solvent-based cleaning fluids. In this case, such items should be stored in a metal container with a close-fitting lid and disposed of on a daily basis.
(e) Dangerous substances should be handled in a well-ventilated area.
(f) Any ill-effects experienced by staff following the use of substances should be reported immediately.

It should be appreciated that the COSHH Regulations apply to offices and to hazardous substances used in offices.

Office equipment

Certain items of office equipment, in particular hand-operated guillotines, can inflict serious injury. Guillotines should be effectively guarded (see Figure 18.1) and staff trained in their safe use.

Other items, such as scissors, letter openers and knives, should only be used for their main purpose and certainly not, for example, as screwdrivers or for opening tins.

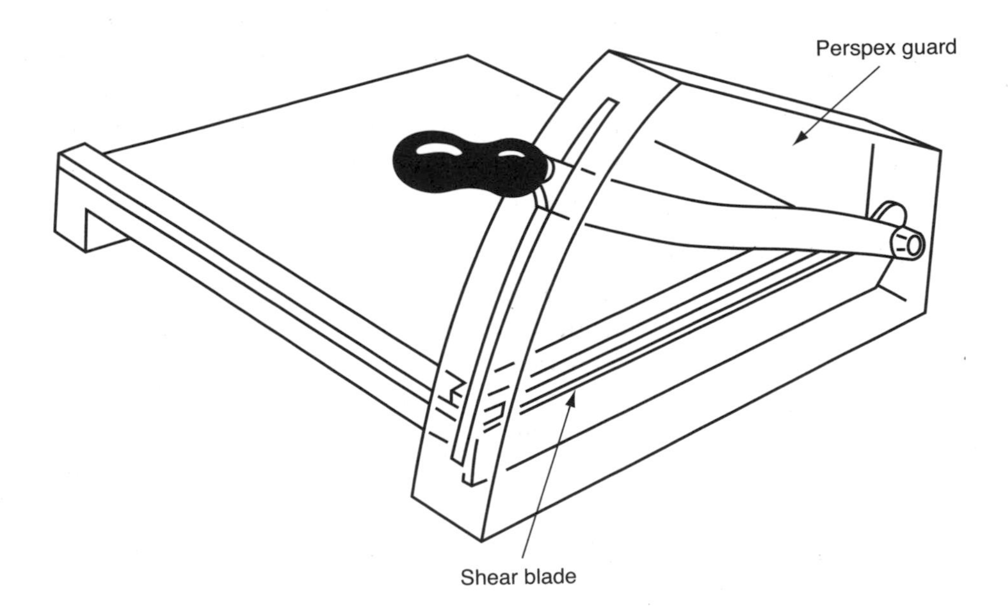

Figure 18.1 Guarding to a hand-operated guillotine

Fire precautions

The majority of fires in offices occur outside normal working hours. The initial cause of the fire will, in many cases, have been created during office hours. To minimise the risk of fire, the following procedures and practices, some of which have already been mentioned earlier in this chapter, should be followed:

(a) On no account should fire exits and designated escape routes be obstructed.
(b) All flammable wastes should be carefully controlled in terms of storage and removal.
(c) The use of radiant-type electric fires should be prohibited.
(d) The requirements of the Health Act 2006 with respect to the ban on smoking in workplaces should be enforced by employers. Where appropriate, external smoking shelters may be provided.
(e) All electrical equipment should be disconnected from the socket when not in use or left unattended for long periods.
(f) Clothing and other items should not be dried close to a direct heat source. In particular, they should not be placed over space heaters.
(g) Prior to locking the office at the termination of work, a trained person should undertake an inspection to ensure that no fire hazards have been created. In larger offices, designated fire wardens should carry out this task.
(h) Flammable items should be kept to a minimum for general use. A metal storage cupboard should be provided for storage.
(i) All staff should be aware of the nearest fire alarm point, the nearest fire appliance and the emergency evacuation plan for the building, including assembly points.

Personal conduct

It is regrettable that many office workers do not envisage their workplace as being potentially dangerous compared, say, with a factory. This attitude can result in accidents, and staff should be trained in the following basic aspects of personal conduct at the induction training stage:

(a) Dress sensibly for work. Do not wear items that may catch in office machinery and do not wear unsuitable footwear.
(b) Look where you are going! Do not read while walking or carry items at such a level that you cannot see where you are going.
(c) Do not run or turn corners quickly when you cannot see what is round the corner.
(d) Use the access equipment provided when storing items normally out of reach, and not revolving typist chairs or existing shelves.
(e) Open doors carefully! There may be someone standing on the other side.

OFFICE HEALTH AND SAFETY AUDIT

To ensure maximum standards of health and safety in the office, the audit shown below should be used:

1. Have the premises been registered under the Offices, Shops and Railway Premises Act 1963?
2. Are the general fire precautions under the Regulatory Reform (Fire Safety) Order being followed and maintained?
3. Has the HSE poster *Health and Safety Law – What you should know* been displayed prominently in the office, in accordance with the Health and Safety (Information to Employees) Regulations 1999? Have leaflets been provided to staff?
4. Are all furniture, fittings and furnishings in a clean state?
5. Is there sufficient space for people to work, bearing in mind the overcrowding standards quoted in the Workplace Regulations?
6. Is the temperature reasonable?
7. Is the means of heating safe and without risks to health?
8. Is a thermometer provided in a conspicuous place on each working floor of the premises?
9. Is ventilation adequate?
10. Is the lighting adequate and in accordance with standards established by the HSE Guidance Note 'Lighting at work'.
11. Are enough toilets provided for both male and female staff?
12. Are there suitable and sufficient washing facilities for both male and female staff, together with soap and drying facilities?
13. Is there an adequate supply of wholesome drinking water?
14. Is there suitable accommodation for external clothing not worn in the office?
15. Are there adequate and suitable seating arrangements?
16. Are all floors, passageways and staircases in a good state of repair and free from obstruction?
17. Do staircases have handrails?
18. Have dangerous parts of machinery been guarded?
19. Is there a suitable number of fully equipped first aid boxes?
20. Are there enough trained first-aiders available during normal working hours?
21. Is the emergency evacuation procedure established and displayed?
22. Are the roll call lists up to date?
23. Are fire precautions records being maintained?
24. Are all fire appliances correctly located and regularly serviced?
25. Are all areas free from accumulations of combustible materials?
26. Are all fire exit doors clearly marked, unobstructed and operational?
27. Can all escape route doors be opened easily?

28. Are all fire-resisting or smoke-stop doors closed and fitted with appropriate door closing gear?
29. Can the fire alarm be heard in all parts of the premises?
30. Has the fire alarm been tested recently?
31. Has a fire drill been carried out in the last 12 months?
32. Have persons been trained in the correct use of fire appliances?

WORKSHOP SAFETY

Many hazards can occur in workshops arising from congestion, inadequate storage, fire, machinery and structural risks, poor environmental conditions and the design and layout of the workshop.

Factors for consideration in assessing the safety of workshops are indicated below.

Structural features

1. **Floors** – clean, sound finish; adequate floor drainage where necessary.
2. **Inspection pits** – safe access; clean; intrinsic flameproof lighting; pit covers / boards or secure fencing when not in use; drainage with sump or pump.
3. **Elevated storage areas** – safe access; adequate lighting; safety rails (minimum 1 metre high) with intermediate rails and toe-boards.
4. **Fixed ladders** – sound; back rings fitted from 2 metres upwards; adequate lighting.
5. **Portable ladders** – sound styles and rung-and-style connections; inspection system for ladders; correct storage.
6. **External areas** – lighting; drainage; yard surfacing; marking out; directional signs; traffic hazards; segregation of pedestrian and traffic routes; parking arrangements; storage of waste.

Environmental aspects

1. **Temperature control** – adequate; minimum 16°C after first hour of working; wall-mounted thermometer; fumes from heating appliances; use of portable heating appliances.
2. **Lighting** – suitable and sufficient; specific lighting arrangements, eg work benches; windows and roof lights clean.
3. **Ventilation** – adequate; LEV systems installed, eg welding operations.
4. **Noise** – machinery and appliances; physical separation; noise reduction; ear protection provided and worn; notices displayed.

Machinery and equipment

1. **Abrasive wheels** – correctly mounted; maximum spindle speed notice displayed (variable speed grinders); tool rest/wheel – maximum 0.125 inches clear; motor isolation when changing wheel; fixed guard; eye protection provided and used; mounting instructions displayed; adequate bench and specific lighting.
2. **Lifting tackle** – SWL marked; inspection and certification every six months; stored off floor; slinging table displayed.
3. **Drills and drilling machines** – flexes and plugs; chuck and spindle guard.
4. **Gas welding** – welding screen/curtain; cylinder racks/chains; full face protection (tinted); cylinder carrier/cradle; hoses; correct cylinder storage.
5. **Electric welding** – welding screens/curtain; cables and clips; equipment and workpiece earth bonded; full face protection (tinted); gloves and apron.
6. **Air compressors** – belt drive guarded; isolation to motor; SWP marked on reservoir; calibrated safety valve; regular draining of reservoir; examination and certification procedure.
7. **Compressed air equipment** – eye protection provided and worn; abuse/misuse; metering valve and tyre pressure gauge; storage of compressed air tools; tyre inflation cage.
8. **Jacking operations** – saddles clean and sound; examined monthly; correctly stored; hydraulic jacks – examination and certification.
9. **Woodworking machinery** – circular saws – guarding below saw table, crown and adjustable guard fitted; use of push sticks; stopping device.
10. **Lathes** – chuck and face plate guard; clamp; eye protection; swarf removal.
11. **Guillotines** – sound construction; fixed or adjustable guard.
12. **Vehicle lifts** – sound construction; wheel stops fitted and operational; chocks provided; warning device/alarm during operation.
13. **Valve grinders** – sound construction; eye protection.
14. **Ramps** – sound construction; wheel stops fitted and used.
15. **Vehicle washing** – drainage; winter icing arrangements; chemical storage and dosing.

Hand tools

1. **Files** – condition; chips and cracks; handles in sound condition; training in correct use.
2. **Hacksaws** – condition; state of blade; training in correct use.

Electrical safety

1. **Hand tools** – low voltage (110v 24v); efficient switches; earthing; double insulation; flexes sound; connections; socket overloading.

2. **Hand lamps** – bulb cage; earthed / low voltage; overhead drop leads; flexes and connections sound.
3. **Battery charging** – flexes and connections sound; earth clamp; adequate ventilation of charging area.
4. **Machine controls** – mushroom headed STOP buttons; inset / shrouded START buttons; clearly identified; isolation.
5. **Control boxes** – kept shut; ON-OFF switch clearly marked.

Fire prevention and control

1. **Flammable materials** – identified; separate storage; potentially flammable atmospheres; spillage control; notice displayed.
2. **Welding equipment** – sitting away from combustible / flammable substances; purging of pits (compressed air).
3. **Flammable wastes** – storage and disposal; incineration hazards.
4. **Diesel store** – tank sound; pump / dispenser sound.
5. **Fire protection and appliances** – fire appliances wall mounted, serviced annually; training in correct use of appliances; fire exits and escape routes correctly marked; key in box / crash bar; exits checked regularly; fire alarm tested regularly; fire drills annually at least.

Hazardous substances

1. **Cleaning and degreasing agents** – dermatitis risks; use of gloves, eye protection, apron; emergency eye / face wash facility; storage tank sound; no smoking; adequate ventilation; labelling of containers and transfer containers.
2. **Solvents** – general awareness of relative flammability and toxicity; storage in accurately labelled containers.
3. **Anti-freeze** – general awareness of relative flammability; storage in accurately labelled containers.
4. **General storage** – ventilated; floor drainage where appropriate; spillage control; personal protection requirements; limited access.

Storage areas

1. **Racking** – stable – fixed to floor / wall; configuration; evidence of overloading; spring racks secure.
2. **Separation** – flammable / dangerous substances stored separately; metal cabinets for small quantities.

Amenity area

1. **Sanitation** – clean; adequate facilities; decoration.
2. **Washing facilities** – clean; hot and cold water; soap, nailbrushes, drying facility; barrier cream in dispenser.
3. **Clothing storage** – adequate; drying facilities for clothing; secure lockers; separation of soiled protective clothing from external clothing where appropriate.
4. **Drinking water** – adequate facilities; fountains.
5. **Mess room/rest room** – adequate; seats provided; ventilated and heated; cooking/food heating facility; sink with hot and cold water; utensil drying facility.
6. **First aid** – first aid box provided and maintained; first-aiders sufficient and trained.

Administration

1. **Registers etc** – planned maintenance record; hoists and slings; abrasive wheels; ladders; power washer maintenance record; high pressure pumps service records; accident book.
2. **Statement of health and safety policy** – clearly displayed.
3. **Abstracts** – 'Health and Safety Law – What you should know' poster displayed.

CATERING SAFETY

Catering activities can be dangerous, particularly where kitchens may be badly designed and staff untrained in the hazards and precautions necessary.

Catering injuries

Typical injuries associated with catering activities include:

(a) scalds to the hands, forearms, feet, legs and trunk through contact with boiling water, hot fats and liquids;
(b) cuts to the hands through the use of knives, slicing machinery and can-opening operations, together with those caused through contact with broken glass and crockery;
(c) burns to the hands and forearms from ovens, hotplates, oven-ware, plates and hot liquids;
(d) bruising, abrasions and fractures due to slips, trips and falls on wet and greasy floors; and
(e) back injuries through incorrect manual handling techniques.

The following types of accidents and injuries to catering staff are common:

Falls on stairs	23%
Falls on kitchen floors	20%
Cuts and hand injuries from slicing and mixing machines	19%
Burns and scalds from various sources	11%
Back injuries from lifting heavy packs	11%
Hand injuries from the use of knives	8%
Falls from step ladders	5%
Strains, sprains and back injuries from moving furniture and equipment	3%

The precautions necessary

1. *Floors*

Falls on kitchen floors can result in broken limbs, head injuries, strains and sprains. It is, therefore, essential that catering staff wear sensible low-heeled shoes with non-slip soles. The floor should be adequately drained, and cleaning and housekeeping procedures should ensure that spillages of fats and liquids are cleared directly after they occur.

2. *Hand tools*

And what about those nasty cuts from boning knives, meat cleavers and saws used in the preparation of meat dishes? Even sharpening a knife using a butcher's steel can be dangerous, particularly if the steel does not incorporate a guard between the handle and the steel. Staff training in the correct use of knives and sharpening procedures is essential.

3. *Machinery*

Catering machinery can include slicing machines, mechanical potato chippers, bowl mixers, waste disposal units and dish washing machines. All these items incorporate moving parts which should be effectively guarded or, alternatively, fitted with appropriate safety devices which prevent the hands coming into contact with moving parts in compliance with the PUWER 1998. Over one-third of machinery-related accidents are associated with slicing machines and are caused by unsafe operation and cleaning of these machines. Causes include inadequate guarding of the machine, removal of the guards while the machine is operating, use of the hand to push forward the meat being sliced, unsafe methods of cleaning the blade, and lack of training in the safe use of such machinery.

4. Environmental factors

A significant indirect cause of accidents in kitchens is frequently the poor levels of working environment provided. Temperatures in kitchens frequently reach 90°F 32°C, many are inadequately ventilated and the high levels of humidity from steaming appliances and boiling liquids can result in staff becoming lethargic and careless. Controlled ventilation giving 12 to 20 air changes per hour, particularly during summer months, together with exhaust ventilation over ranges is essential. Well-designed lighting systems in preparation, storage and external areas should be provided and maintained.

5. Fire

The risk of fire in catering activities needs urgent consideration due to the very nature of much of the equipment used – ovens, frying equipment, grills and other forms of open-flame appliance. Many food ingredients and cooking aids are highly flammable, such as cooking oil, butter, margarine and lard. They spread rapidly on a floor taking a fire to other parts of a kitchen. In particular, staff should be trained to deal with small fat fires on cooking ranges, there should be an adequate number of fire appliances and fire blankets available, fire exits should be marked and kept clear, and regular fire drills held. It is vital that staff should know the right type of appliance to use with different classes of fire.

6. Manual handling

Catering staff are frequently required to handle heavy weights, such as meat joints, turkeys, packs of dry goods, sacks of vegetables and heavy utensils. A substantial number of injuries are caused by incorrect manual handling or through staff attempting to lift items which are too heavy. With any manual handling operation, the rule must be 'If you can't lift it, get help'.

One of the principal causes of back injury is lifting heavy loads from the bottom of a chest freezer where it is virtually impossible to practise the principles of safe lifting. Wherever possible, heavy frozen items, such as meat joints, turkeys and bagged vegetables should be stored in purpose-built cold stores or upright freezer units to eliminate the risk of back injury.

7. Personal protective clothing

Many accidents to catering staff are caused through the incorrect use of personal protective clothing – long sleeves, open coats, inadequate hair covering. For both safety and hygiene reasons, protective clothing should be reasonably tight-fitting and the hair completely covered by a net. This includes male catering staff!

8. Jewellery

Significant injuries to fingers are associated with the use of rings. Only a plain wedding ring should be worn.

9. ***Hazardous substances***

A wide range of hazardous substances are encountered in catering operations and, in some cases, staff are inadequately trained and supervised in their correct use. Detergents and detergent sanitisers may be strongly acidic or alkaline. Used at an incorrect dilution, they can cause skin damage in the form of rashes to the hands and arms and, in some cases, dermatitis. Dilution instructions must, therefore, be carefully followed. Staff engaged in the use of hazardous cleaning substances should be provided with elbow-length gloves and be prevailed upon to use them whenever there is a risk of hand contact with such substances. In some cases it may be necessary to undertake health risk assessments of substances hazardous to health under the COSHH Regulations 2002.

SUMMARY

1. Most people consider office work to be relatively safe and do not appreciate the risks, particularly that of fire.
2. Poor standards of housekeeping are one of the principal causes of accidents in offices.
3. Many items of office furniture and fittings can be dangerous if not used properly.
4. Abuse and misuse of electrical appliances and equipment can result in fires and the risk of electrocution.
5. All staff should be trained in correct manual-handling techniques, and handling equipment should always be provided to reduce the risk of handling accidents.
6. Staff should be aware of the dangerous substances in use and the precautions necessary.
7. A clearly established fire procedure is essential.
8. Unsafe behaviour by office staff should not be tolerated, with disciplinary action being taken against offenders.
9. An office safety audit should be undertaken at six-monthly intervals by a responsible person.
10. The principal hazards in workshops are associated with congestion, poor housekeeping, inadequate lighting, various types of machinery and equipment, storage of flammable substances and electricity.
11. Workshops are subject to the requirements of the Workplace Regulations and the Provision and Use of Work Equipment Regulations 1998.

12. Storage areas should be fitted with suitable racking which should be well maintained.
13. The principal types of accident in catering operations are falls, cuts and hand injuries.
14. Kitchen floors can be extremely dangerous if poorly maintained and not cleaned regularly.
15. A wide range of hazardous substances is encountered in catering operations, in various forms of detergent.

CONCLUSION TO PART 4

Safety technology covers an extremely wide field, and a broad understanding of many engineering disciplines is essential in order to comply with the legal requirements covering machinery, the safe use of electricity, fire protection, mechanical handling and construction activities.

There are still far too many accidents associated with unguarded machinery. Many of these accidents have fatal consequences or can result in people being maimed for life. Application of machinery safety principles detailed in BS EN ISO 12100 'Safety of machinery' is, therefore, essential if future accidents associated with machinery operation are to be avoided.

The costs to the nation of fire incidents, including those associated with deaths, property damage and lost production, have been substantial in the last decade. Organisations need to examine their fire protection procedures on a regular basis, including enlisting the assistance of the fire authorities wherever guidance is needed. It only needs one major fire to put a company out of business.

Over 1,500 people are killed or injured at work each year as a result of unsafe practices when using electricity or unsafe electrical installations. All staff should be trained in the principles of electrical safety and in safe working practices when using same.

Increasing attention is being paid by the enforcement agencies to the need for a safe working environment, including the elimination of the traditional 'sweat shops' associated with a number of industries. Environmental stress, associated with poor levels of temperature and humidity control, lighting and ventilation, is common in many workplaces. In many industries, high noise levels are accepted by management and workers as an intrinsic feature of that industry, and claims for occupational deafness are a standard business cost. This state of affairs cannot be tolerated and companies should place much greater emphasis on the need to reduce or eliminate all forms of environmental stress. The benefits, as with other areas of health and safety improvement, can be substantial, including reduced absenteeism, reduced health-related claims and improved operator performance.

REFERENCES

Health and Safety Commission (1992) *Workplace Health, Safety and Welfare Regulations 1992 and Approved Code of Practice*, HMSO, London

Health and Safety Executive (1987) *Catering Safety: Food preparation machinery*, HMSO, London

Health and Safety Executive (1987) *Health and Safety in Kitchens and Food Preparation Areas*, HMSO, London

Health and Safety Executive (1991) *Health and Safety in Motor Vehicle Repair HS(G)67*, HMSO, London

Royal Society for the Prevention of Accidents (1972) *Health and Safety in Offices and Shops*, RoSPA, Birmingham

Royal Society for the Prevention of Accidents (1976) *Catering Care*, RoSPA, Birmingham

Index